The Mathematical World of
Charles L. Dodgson (Lewis Carroll)

Charles Lutwidge Dodgson (1832–1898)

THE MATHEMATICAL WORLD OF CHARLES L. DODGSON (LEWIS CARROLL)

Edited by

ROBIN WILSON

AMIROUCHE MOKTEFI

OXFORD

UNIVERSITY PRESS

OXFORD
UNIVERSITY PRESS

Great Clarendon Street, Oxford, OX2 6DP,
United Kingdom

Oxford University Press is a department of the University of Oxford.
It furthers the University's objective of excellence in research, scholarship,
and education by publishing worldwide. Oxford is a registered trade mark of
Oxford University Press in the UK and in certain other countries

© Robin Wilson and Amirouche Moktefi 2019

The moral rights of the authors have been asserted

First Edition published in 2019

Impression: 1

Published in the United States of America by Oxford University Press
198 Madison Avenue, New York, NY 10016, United States of America

British Library Cataloguing in Publication Data
Data available

Library of Congress Control Number: 2018961475

ISBN 978–0–19–881700–0

DOI: 10.1093/oso/9780198817000.001.0001

Printed and bound by
CPI Group (UK) Ltd, Croydon, CR0 4YY

CONTENTS

PREFACE

Scholarship on Charles Lutwidge Dodgson has long focused on his literary works, and specifically on the *Alice* tales, *Alice's Adventures in Wonderland* and *Through the Looking-Glass*. Theories have abounded on Dodgson's dual personality: the fantastic novelist on the one hand and the dull mathematician on the other. The latter was generally ignored, or even dismissed, except for specific occasions such as the 1932 centenary of his birth which witnessed the publication of a few commentaries on Dodgson's mathematics, notably by R. B. Braithwaite and D. B. Eperson. Yet, Dodgson was first and foremost studied as the 'man who wrote *Alice*'.

The situation progressively changed in the mid-20th century. This was stimulated by, among other reasons, the abridged edition of Dodgson's diaries in 1953, the new editions of some of Dodgson's mathematical works by Dover Publications in 1958 and later, and the appearance of serious studies by Helmut Gersheim on Dodgson's photography and Duncan Black on Dodgson's political papers. These studies contributed to shifting Dodgson's image towards a unique yet multifaceted character with numerous interests. Although the novelist continued to dominate scholarship, studies on Dodgson's mathematics and photography episodically appeared. In 1954 Warren Weaver produced the first description of Dodgson's mathematical *Nachlass*.

This revival of Dodgson studies was accentuated in the 1960s, and more decisively in the 1970s, with the publication of Martin Gardner's *Annotated Alice* in 1960 and the foundation of Lewis Carroll Societies in Britain (1969) and North America (1974). These societies (and others) gave scholars opportunities to meet regularly and to publish their findings in the societies' journals and other publications. Studies on Dodgson's mathematics began to appear on a regular basis.

By the 1980s it had become evident that Dodgson's studies were no longer exclusively Carrollian. This was seen in the larger space devoted to Dodgson's mathematics in the new biographies, in collections of essays, and finally in meetings and occasional celebra-tions. Significant studies were published by Tony Beale, William Warren Bartley, III, Ernest Coumet, Edward Wakeling, George Englebretsen, Eugene Seneta, Iain McLean, Mark R. Richards, Francine F. Abeles, and others. Dodgson's mathematics became a serious academic subject for historians of mathematics.

In the early 1990s two major editorial projects were initiated, supported by the Carrollian societies. These projects were rightly expected to make a significant impact on Dodgson scholarship in general, and on the study of his mathematical activities in par-ticular. First, Edward Wakeling engaged in editing an unabridged version of Dodgson's diaries, thus restoring the mathematical entries that had been largely omitted from earlier editions. A series of collections of Dodgson's pamphlets was then projected, of which five volumes have so far appeared, under the editorship of Edward Wakeling (on Oxford affairs), Francine Abeles (mathematics, political issues, and logic), and Christopher Morgan (games and puzzles). There is no doubt that these publications have stimulated further research into several areas of Dodgson scholarship, and notably on his mathematics.

In the 21st century it has not been unusual to encounter studies on Dodgson's mathematics at international conferences or in academic journals. Previously unknown sources have been revealed, novel results have been discovered, and new scholars have investigated a range of Dodgson's mathematical activities. In recent years, spe-cific meetings have even been devoted entirely to Dodgson's mathematics, and ensu-ing collections of essays have been published. In 2008 one of us wrote the first mathematical biography of Dodgson, aiming to fill a need that had long been felt, and to make accessible to a wide audience a large amount of information that had hitherto been unreachable.

The present volume might be viewed as a culmination of this long line of work. It gathers the best authorities on Dodgson's mathematics in their areas of expertise. It reveals Dodgson as a regular, obtuse, and imaginative mathematical researcher. His interests cover a variety of mathematical disciplines. He addressed problems that may be viewed as highly abstract, but also investigated topics that would rather be ranked among applied mathematics. Finally, it depicts someone who shared many of the concerns of his contemporary British mathematicians but was less familiar with (although not totally ignorant of) the progress that was being made on the Continent. It may be said of Dodgson that he truly was a Victorian mathematician.

We wish to express our gratitude to the many people who have helped us in the preparation of this book. In particular, we should like to thank all the contributors to this book for their patience, during the editing process, in coping with our many whims and requests. In particular, we are very grateful to Edward Wakeling for suggesting and providing many of the pictorial images in this volume from his magnificent archive collection. On the publication side, we should like to thank Daniel Taber and Katherine Ward of Oxford University Press and Lydia Shinoj of SPi Global for all their help and encouragement.

Robin Wilson and Amirouche Moktefi

June 2018

CHRONOLOGY OF EVENTS

This is not a full list of Charles Dodgson's activities, but contains the main events in his life and his most important mathematical (and other) publications; several titles are abbreviated.

1832 27 January: Charles Lutwidge Dodgson born at Daresbury, Cheshire
1843 Moves to Croft Rectory, Yorkshire
1844 Attends Richmond School
1846 Attends Rugby School
1849 Returns to Croft Rectory
1850 Matriculates at Oxford University
1851 Comes into residence at Christ Church, Oxford
       Mother dies
1852 Elected a 'Student' at Christ Church
1854 Long Vacation at Whitby studying with Bartholomew Price
       First Class in Mathematics in his Finals Examinations
       Receives Bachelor of Arts degree
1855 Begins teaching at Christ Church
       Henry Liddell appointed Dean of Christ Church
       Elected Mathematical Lecturer at Christ Church
1856 Adopts the pseudonym Lewis Carroll
       Begins hobby of photography
1857 Receives Master of Arts degree
       Hiawatha's Photographing
1860 *A Syllabus of Plane Algebraic Geometry*
       Notes on the First Two Books of Euclid

1884 *The Principles of Parliamentary Representation*
1885 *A Tangled Tale*
1886 *Alice's Adventures Under Ground* (facsimile edition)
1887 *The Game of Logic* (earlier private edition, 1886)
       To Find the Day of the Week for Any Given Date
1888 *Curiosa Mathematica, Part I. A New Theory of Parallels*
       Memoria Technica
1889 *Sylvie and Bruno*
1890 *The Nursery "Alice"*
1892 Resigns as Curator of Christ Church Common Room
1893 *Curiosa Mathematica, Part II. Pillow-Problems*
       Sylvie and Bruno Concluded
1894 *A Logical Paradox*
1895 *What the Tortoise Said to Achilles*
1896 *Symbolic Logic. Part I. Elementary*
1897 *Brief Method of Dividing a Given Number by 9 or 11*
       Abridged Long Division
1898 14 January: Charles Dodgson dies in Guildford

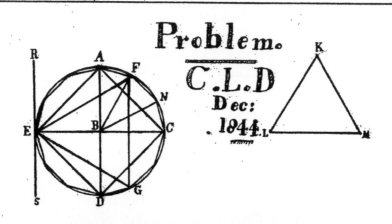

Problem.

C.L.D
Dec: 1844

To trisect a right angle, that is, to divide it into three equal parts.

Let there be a right angle ABC, it is required to trisect it.

Produce AB to D and make BD equal to AB, and make BE equal to AB and produce CB to E and make EB equal to BC, and join AE, ED, DC, CA. Because AB is equal to BD, and BE is common to the two triangles ABE, DBE, and the angle ABE is equal to the angle DBE, therefore the base AE is equal to the base ED; and in like manner it may be proved that all the four AE, ED, DC, CA are equal, therefore AEDC is equilateral, and because the three angles of a triangle are equal to two right angles, and that the angle ABE is a right angle, (for ABC is a right angle, and EC is a straight line) therefore the angles BAE, BEA are equal to one right angle and because BA is equal to BE, therefore the angle BAE is ½ a right angle, and in like manner it may be proved that the angle BAC is ½ a right angle, therefore the angle BAC is a right angle, and in like manner it may be proved that the angles AED, EDC, DCA are also right angles, therefore AEDC has all its angles right angles, and it was proved

A page of Charles Dodgson's geometry, written when he was aged 12

[*The Colophon*, New Graphic Series No. 2, New York, 1939]

CHAPTER 1

A mathematical life

ROBIN WILSON AND AMIROUCHE MOKTEFI

Early years

Charles Lutwidge Dodgson was born on 27 January 1832 at the Old Parsonage at Daresbury in Cheshire.

His father, the Reverend Charles Dodgson, came from a long line of clergy stretching back several generations. A deeply religious man with a passionate interest in mathematics, he enjoyed a brilliant early career at Oxford University where he received a double First Class degree in Classics and Mathematics at Christ Church in 1821. A Studentship at Christ Church (somewhat equivalent to a Fellowship in other colleges) entitled him to live in college for the rest of his life, provided that he remained unmarried and prepared for holy orders. He was ordained Deacon in 1823 and Priest in the following year.

In 1827 he married Fanny Lutwidge, his first cousin, and duly forfeited his Studentship. Christ Church presented him with a living at the parish church in Daresbury, where he and Fanny started their large family of seven girls and four boys. All survived to adulthood, and Charles, as the eldest son, soon established himself as their natural leader, delighting in entertaining his younger brothers and sisters.

The Dodgson family received a strict Christian upbringing, with Sunday devoted solely to reading religious books, learning extended passages from the Bible, and attending services at the Church for their father's extempore sermons. Charles inherited a deep religious conviction that would govern his future life.

The Mathematical World of Charles L. Dodgson (Lewis Carroll). Robin Wilson and Amirouche Moktefi.
Oxford University Press (2019). © Oxford University Press 2019.
DOI: 10.1093/oso/9780198817000.001.0001

The Revd. Charles Dodgson

At Daresbury the Reverend Dodgson's meagre stipend of less than £200 per year required the parents to educate their children at home. Charles, in particular, received from his father a thorough grounding in mathematics, Latin, Christian theology, and English literature, subjects that would feature prominently throughout his life. Of his mathematical precocity, the story is told that:[1]

One day, when Charles was a very small boy, he came up to his father and showed him a book of logarithms, with the request, "Please explain." Mr. Dodgson told him that he was much too young to understand anything about such a difficult subject. The child listened to what his father said, and appeared to think it irrelevant, for he still insisted, "*But*, please, explain!"

In 1843 the Dodgson family moved to Croft-on-Tees where his father became the rector of the parish church. At Croft, Charles enjoyed an idyllic childhood with his brothers and sisters, with delightful walks in the Yorkshire countryside and many games to play. He derived much pleasure from writing and painting and entertaining the family with puppet shows and conjuring displays. During one winter he constructed a maze in the snow 'of such hopeless intricacy as almost to put its famous rival at Hampton Court in the shade'.[2]

With his move to Croft, the Reverend Dodgson's income increased considerably, and he could now afford to send Charles to a private school to build his son's character and to prepare him for a career in the Church. In August 1844 Charles went to the Free Grammar School in Richmond, a school of 120 pupils just ten miles from Croft, where the curriculum consisted mainly of religious instruction and the classical languages and

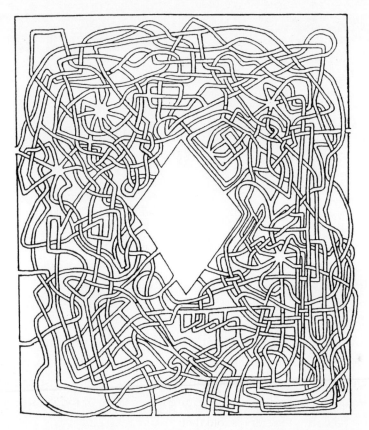

A maze constructed by Charles Dodgson for the family magazine *Mischmasch* around 1855: it is said to be the oldest example of a three-dimensional maze

literature, with mathematics, French, and accounting as optional extras. The arithmetic textbook used there was the 1842 edition of Francis Walkingame's classic 18th-century text, *The Tutor's Assistant*, which contained such questions as:[3]

What is the cube root of 673373097125? (*Ans.* 8765)
If from London to York be accounted 50 leagues, I demand how many miles, yards, feet, inches, and barley-corns? [A league is 3 miles; a barleycorn is $\frac{1}{3}$ inch.]
(*Ans.* 150 *miles*, 264000 *yards*, 792000 *feet*, 9504000 *inches*, 28512000 *barleycorns*.)
If 504 Flemish ells, 2 qrs. cost 283 l. 17s. 6d.; at what rate must I give for 14 yards?
(*Ans.* £10:10s)

By this time Charles was already composing Latin verse, and a page of geometry, written at age 12, shows how far his mathematical interests and abilities were developing

The Free Grammar School, Richmond, and Rugby School

(see the picture that introduces this chapter[4]): it is one of the earliest examples of his handwriting.

In February 1846 the 14-year-old Charles Dodgson was sent to Rugby School, where he delighted in his studies of mathematics and the Classics but was subjected to bullying. His health also suffered: in 1848 he developed whooping cough and later contracted mumps, aggravating the deafness in his right ear that had developed some years earlier. In later years Charles would look back on his Rugby days with distaste.

The teaching and curriculum at Rugby were traditional. Each week the instruction, which started at 7 a.m., consisted of sixteen lengthy classes in the Classics, history, and Scripture, compared with only two classes in French, two in mathematics, and none in science. In spite of this, Charles made good progress with his studies, being considered exceptionally gifted in mathematics and winning prizes for history, divinity, mathematics, Latin composition, and English. At the end of his first year he came first in mathematics, a year later he won the 2nd Mathematical prize in the annual general mathematics examination, and around Christmas 1848 he achieved 1st class in mathematics and other subjects.

His teachers were enthusiastic about his progress. In 1848 his mathematics master Robert Mayor confided to the Reverend Dodgson that he 'had not had a more promising boy at his age since he came to Rugby', and just before Charles left Rugby School the headmaster, Dr Archibald Tait (later Archbishop of Canterbury), wrote:[5]

I must not allow your son to leave school without expressing to you the very high opinion I entertain of him…His mathematical knowledge is great for his age, and I doubt not he will do himself credit in classics. As I believe I mentioned to you before, his examination for the Divinity prize was one of the most creditable exhibitions I have ever seen.

After the hardships of Rugby School Charles returned to Croft Rectory for a few months to prepare himself for the next stage of his career, as he put his schoolboy difficulties behind him and headed for Oxford.

An undergraduate at Oxford

With his impressive Rugby School record the young Charles Dodgson was well placed to study at Oxford University, and on 23 May 1850 he travelled to Oxford for his matriculation examinations in Latin, Greek, and mathematics, and for the ensuing ceremony in which he pledged allegiance to the thirty-nine articles of the Church of England and was officially enrolled as a member of the University.

At that time Oxford was a small country town with unpaved streets and horse-drawn carriages. Then, as now, the University consisted of a number of colleges, and Charles became a member of Christ Church, then the largest college, where his father had achieved his great successes thirty years earlier.

Founded in 1546 by Cardinal Wolsey, King Henry VIII's Chancellor, Christ Church includes the Cathedral of the Diocese of Oxford. During the English Civil War of the 1640s King Charles I lived in Christ Church and held Parliamentary meetings in the magnificent dining hall (known to recent generations of Harry Potter enthusiasts as 'Hogwarts Hall'), where Dodgson later claimed to have dined many thousands of times. Every evening Great Tom, the bell in Christopher Wren's gate-tower, still rings 101 times to celebrate the 101 Students who became part of the Foundation in the 17th century.

Christ Church, Oxford, in 1850

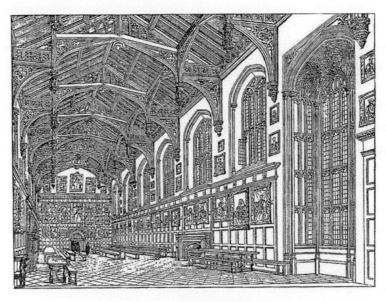

The Dining Hall of Christ Church

When Charles Dodgson entered Christ Church for the first time, at the age of 18, he could hardly have expected that this would remain his home for the rest of his life. However, he did not immediately take up residence, but returned to Croft Rectory to prepare himself for the start of his course.

Charles's return to Oxford on 24 January 1851 was short-lived. Two days later his beloved mother Fanny died suddenly and unexpectedly of inflammation of the brain at the age of only 47, and he had to return home. This event was devastating for Charles, and particularly for his father who needed to arrange for the care of his large family. After a short period Fanny's sister, Lucy Lutwidge, arrived at Croft Rectory to care for the Dodgson children; Aunt Lucy remained with the family for the rest of her life.

Back in college, Charles settled into the routine of undergraduate life. Called every morning at around 6.15 a.m., he had breakfast in his rooms, attended college chapel at 8 o'clock, and studied throughout the morning, wearing his gown and mortar board for lectures and tutorials and around the town. In the afternoons he relaxed – going for long walks, boating with friends on the river, or watching a game of cricket. After dinner in the Great Hall at 5 p.m. he often spent his evenings reading or composing letters, standing at his writing desk. Although several Christ Church undergraduates were noblemen from wealthy families who spent their time in riotous living, such as hunting, gaming, and drinking, Charles, like his father thirty years earlier, was there for the purpose of serious study and the passing of examinations.

The University year was divided into four terms. Charles's first term was Hilary (or Lent) Term, from January until the end of March. This was followed by two short terms, Easter Term from late April to early June, and Trinity (or Act) Term from mid-June to early July. The Long Vacation extended for three months, to be followed by Michaelmas Term, which lasted from October until December.

Undergraduates could choose to be *passmen*, working for a Pass degree that took about three years, or *classmen*, working for an Honours degree that took a year longer. For the Honours degree, students had to pass in two subjects: the first was required to be Literae Humaniores (Classics) and the second could be selected from Mathematics, Natural Science, or Law and Modern History. Charles elected to be a classman working towards the four-year Honours degree in Classics and Mathematics.

At Oxford the teaching was carried out in University lectures given by the professors, and by college lecturers who lectured to small groups of undergraduates and tutors who gave private tuition. Lectures in pure mathematics were given by the Savilian Professor of Geometry, the Revd Baden Powell, while at Christ Church the Mathematical Lecturer was the newly appointed Robert Faussett, who taught Dodgson throughout his undergraduate career and became a close friend.

Baden Powell, Savilian Professor of Geometry

Robert Faussett, Lecturer at Christ Church

[Senior Common Room of Christ Church, Oxford]

Charles Dodgson's university course required him to take three main examinations involving written papers and viva voce confrontations – Responsions, Moderations, and Finals.

Responsions (colloquially known as 'Little-go') was the first hurdle that all students had to overcome. It took place twice a year and consisted of papers on the Latin and Greek authors, a translation from English into Latin, a paper of grammatical questions, basic arithmetic (up to the extraction of square roots), and a paper on algebra or geometry. Most students attempted Responsions after a year or more, but young Dodgson was better prepared than most and took it in Trinity Term 1851.

After spending the long summer vacation of 1851 back at Croft Rectory, Dodgson returned to Oxford to begin preparations for the second part of his examinations a year later, attending lectures on pure geometry by Baden Powell and continuing to study with the college lecturer. For his continued progress the college awarded him a scholarship of £20 per year.

The 'First Public Examination under Moderators' took place in November 1852. This consisted of papers on the four Gospels in Greek, Greek and Latin authors, and either a logic paper or one in geometry and algebra.[6] In addition, candidates for Honorary Distinction in *Disciplinis Mathematicis* were required to take a paper in pure mathematics. Dodgson achieved a First Class in Mathematics and a Second Class in Classics.

An Oxford University viva voce examination

A college Studentship for his son had long been an ambition of the Revd Charles Dodgson. As a result of young Charles's success in his Moderations exams, he became a Student of Christ Church (as his father had been a generation earlier). This entitled him to reside in college, provided that he remained celibate and prepared for holy orders, conditions that he took very seriously. His father was delighted, writing to Charles:[7]

The feelings of thankfulness and delight with which I have read your letter just received, I must leave to *your conception*; for they are, I assure you, beyond *my expression*; and your affectionate heart will derive no small addition of joy from thinking of the joy which you have occasioned to me, and to all the circle of your home.

It was indeed a busy time for the Reverend Dodgson. In late 1852 he had been appointed a Canon of Ripon Cathedral in Yorkshire, requiring him to spend the first three months of each year in Ripon before returning home to Croft Rectory for the other nine months. Soon afterwards he would also be appointed Archdeacon of Richmond. He continued to take an interest in his son's mathematical activities for the rest of his life.

The Public Examination of *Finals* was the culmination of Charles's undergraduate career, and consisted of two parts. The first of these was *Greats*, in the Easter term of 1854, which tested the Classical languages and literature, together with ancient history and philosophy, and was compulsory for everyone. In spite of working thirteen hours a day for the three weeks before the examination, and spending the whole night before the viva voce examination over his books, Charles emerged with only a Third Class.

A letter on geometric series from the Revd. Charles Dodgson to his son

[Morris L. Parrish Collection, Princeton University]

The second part of his Finals degree was in his chosen area of mathematics, for which the minimum requirement was 'The first six books of Euclid, or the first part of Algebra'. Candidates for Honours were also required to study 'Mixed as well as Pure Mathematics', involving a range of topics from the differential and integral calculus to astronomy and optics.

In order to prepare for these examinations Dodgson spent much of the 1854 summer vacation on a two-month mathematical reading party to Whitby in Yorkshire. This was led by Bartholomew Price, the recently appointed Sedleian Professor of Natural Philosophy and the author of distinguished treatises on the calculus. (His first name was frequently abbreviated to 'Bat', and features in *Alice's Adventures in Wonderland* in the Hatter's parody, 'Twinkle, twinkle, little bat!'.) Dodgson developed a great admiration for Professor Price, who would later become a close friend and adviser. From Whitby he confided to his sister Mary that 'I am doing Integral Calculus with him now, and getting on very swimmingly'.[8]

Charles's Finals examinations in mathematics took place in October and November 1854, and ranged over many areas of pure and applied mathematics. The examiners were Bartholomew Price, William Donkin (the Savilian Professor of Astronomy), and Henry Pritchard of Corpus Christi College, and the examination resulted in a Class list of five First Class degrees, seven Seconds, no Thirds, six Fourths, and thirty-five Pass degrees

SECOND PUBLIC EXAMINATION.

I.

Geometry and Algebra.

1. Compare the advantages of a decimal and of a duo-decimal system of notation in reference to (1) commerce, (2) pure arithmetic; and shew by duodecimals that the area of a room whose length is 29 feet $7\frac{1}{4}$ inches, and breadth is 33 feet $9\frac{1}{4}$ inches, is 704 feet $30\frac{3}{8}$ inches.

2. Planes which are perpendicular to parallel straight lines are parallel to one another: and all planes which cut orthogonally a given circle meet in one and the same straight line.

3. Solve the following equations:

(1) $\dfrac{x + \sqrt{a^2 - x^2}}{x - \sqrt{a^2 - x^2}} = b.$

(2) $\left. \begin{array}{l} x^3 - y^3 = 98 \\ x - y = 2 \end{array} \right\}$

(3) $\left. \begin{array}{l} \dfrac{x}{a} + \dfrac{y}{b} = 1 \\[4pt] \dfrac{z}{c} + \dfrac{x}{a} = 1 \\[4pt] yz = bc \end{array} \right\}$

4. The difference of the squares of any two odd numbers is divisible by 8.

5. Shew that in a binomial, (whose index is a positive whole number,) the coefficient of any term of the expansion reckoned from the end, is the same as the coefficient of the corresponding term reckoned from the beginning.

6. In a given equilateral triangle a circle is inscribed, and then in the triangle formed by a tangent to that circle parallel to any side and the parts of the original triangle cut off by it, another circle is inscribed, and so on *ad infinitum*. Find the sum of the radii of these circles.

[Turn over.

An Oxford University Finals paper in mathematics from 1854

(listed as Fifth Class). The young Charles was extremely successful, coming top of the entire list.[9]

Charles Dodgson received his Bachelor of Arts degree at the graduation ceremony on 18 December 1854, bringing his undergraduate days to a triumphant close.

Oxford lecturer

With his Finals examinations safely behind him, Dodgson returned to Oxford in early 1855 to resume life as a Student at Christ Church. Robert Faussett was about to give up his Mathematical Lectureship and leave Oxford to take up an army commission fighting in the Crimean War.

Every year Oxford University awarded Junior and Senior Mathematical Scholarships by examination, the latter being usually taken after Finals, and Dodgson resolved to study for the Senior Scholarship examinations in March 1855. In order to improve his chances of winning the scholarship he arranged regular coaching from Professor Price, whose company he had so enjoyed during the previous summer's reading party. However, the work proved to be less straightforward than he had expected:[10]

I talked over the "Calculus of Variations" with Price today, but without any effect. I see no prospect of understanding the subject at all.

Charles Dodgson as a young don

Additionally, his attentions were increasingly being directed towards college teaching. He was spending about fifteen hours per week teaching individual pupils, which left little time to prepare for the Senior Scholarship. The inevitable happened: in March the scholarship examinations came and went, and he was unsuccessful. But he was clearly ashamed of his performance:[11]

It is tantalising to think how easily I might have got it, if only I had worked properly during this term, which I fear I must consider as wasted. However, I have now got a year before me…I mean to have read by next time, Integral Calculus, Optics (and theory of light), Astronomy, and higher dynamics.

I record this resolution to shame myself with, in case March /56 finds me still unprepared, knowing how many similar failures there have been in my life already.

As we have seen, a major cause of Dodgson's lack of success in the Scholarship examinations was the time consumed by his teaching commitments. At the beginning of term he had been approached by the Senior Censor of Christ Church to instruct a freshman who was preparing for Responsions:[12]

Had my first interview with Burton, my first pupil: he seems to take in Algebra very readily. I doubt if it will be worth his while to coach two terms merely for his Little-Go – another lesson or two will decide.

Gradually he took on other private pupils, and by Easter term their number had increased dramatically to fourteen. Although not an official arrangement, Dodgson considered that the experience so gained would increase his chances of getting the Mathematical Lectureship, as well as bringing in about £50. He organized his students into tutorial groups, remarking ruefully that the fifteen hours of teaching each week would be 'a remedy against idleness, such as I could never have devised for myself'.[13] The topics that he taught included the differential calculus, conics, trigonometry, Euclidean geometry, and algebra.

Dodgson's mark sheet for five algebra pupils in 1877

[Warren Weaver Collection, University of Texas at Austin]

The Revd Henry Liddell, Dean of Christ Church

[Governing Body of Christ Church, Oxford]

Financially, Dodgson was now managing to stand on his own feet. He had been appointed college Sub-Librarian in February, which brought in £35 per year, and in May he recorded that:[14]

The Dean and Canons have been pleased to give me one of the "Bostock" Scholarships – said to be worth £20 a year – this very nearly raises my income this year to independence. Courage!

In June 1855 the Dean of Christ Church died and his successor was duly chosen by Queen Victoria, as Visitor to the college. This was the Revd Henry Liddell, Headmaster of Westminster School and half of the formidable team of Liddell and Scott who produced the Greek–English Lexicon which is still in use by undergraduates today. The Liddells had four children, one of whom, Alice (then aged 3), would forever be associated with the name of Lewis Carroll.

Over the summer period the new Dean appointed Dodgson to the Mathematical Lectureship which he so desired, to start at the beginning of 1856. Dodgson quickly made a resolution not to try again for the Scholarship.

One of the college traditions on the installation of a new Dean was for the Canons of Christ Church to appoint a 'Master of the House', an arrangement that gave the holder the privileges of a Master of Arts within the college. In October 1855 Dodgson was selected for this honour. He did not receive his official Master of Arts degree from the University until February 1857.

It had been an eventful year for the young Charles Dodgson. On 31 December 1855 he looked back on the past twelve months:[15]

I am sitting alone in my bedroom this last night of the old year, waiting for midnight. It has been the most eventful year of my life: I began it as a poor bachelor student, with no definite plans or expectations; I end it a master and tutor in Ch. Ch., with an income of more than £300 a year, and the course of mathematical tuition marked out by God's providence for at least some years to come. Great mercies, great failings, time lost, talents misapplied – such has been the past year.

An opportunity for some schoolteaching arose in Oxford during the following winter, when he tried his hand at some mathematics teaching at St Aldate's School, directly across the road from Christ Church. Dodgson varied his lessons with stories and puzzles, and he may have been the first to use recreational topics as a vehicle for conveying more serious mathematical ideas. But the pupils increasingly became restless and inattentive, and before long he was relieved to give it up.

Meanwhile, the college lectureship was proving to be burdensome, as Dodgson became increasingly overwhelmed by his teaching commitments. His lectures towards the end of 1856 were occupying seven hours a day, leaving him inadequate time for preparation, and he felt himself daily becoming more and more unfit for the position he held:[16]

I am becoming embarrassed by the duties of the Lectureship, and must take a quiet review of my position, to see what can be done…I have five pupils, whose lectures need preparing for, namely *Blackmore* in for a First at Easter, doing end of Differential Calculus (*new to me*), and to begin Integral Calculus soon.

Rattle in for a First in Mods this time, needs special problems etc. and very probably high Diff: Cal:, a little Int: Cal: and Spherical Trig.

Blore in for a Second, easier problems etc.

Bradshaw in this time next year, reading the circle in *Salmon*, and is already in work new to me.

Harrison in for the Junior Scholarship this term, we are beginning *Salmon*, so that his case is included in Bradshaw's, and he is reading with Price as well, which makes his case easier.

…Something must be done, and done *at once*, or I shall break down altogether.

Teaching for Responsions was also proving to be unrewarding:[17]

I am weary of lecturing and discouraged. I examined six or eight men today who are going in for Little-Go, and hardly one is really fit to go in. It is thankless uphill work, goading unwilling men to learning they have no taste for, to the inevitable neglect of others who really want to get on.

Fortunately, the situation gradually resolved itself as he gained in confidence and experience, and he continued to hold the lectureship for a further twenty-five years.

Dodgson's other lives

It was around this time that Dodgson began to publish his writings and verses. In mid-1855 he had made contact with Edmund Yates, editor of *The Train: A First-Class Magazine*, and offered him poems, parodies, and short stories, several of which duly appeared. Yates suggested that he adopt a pseudonym for his comic writings so as to distinguish them from his academic publications, and Dodgson eventually settled on Lewis Carroll: 'Lewis' from Lutwidge, his middle name and mother's maiden name, and 'Carroll' from Carolus, the Latin for Charles. He used his pen-name when writing books for a general audience, such as his *Alice* books and his books on logic.

Amateur photography had become the popular craze of the 1850s. In September 1855 Dodgson's Uncle Skeffington, always one for the latest gadgets, visited the family at Croft with his camera and Dodgson joined him in some photographic excursions, quickly becoming addicted and remaining so for the next twenty-five years. In March 1856 he travelled to London to purchase a fine rosewood box camera and lens for the substantial sum of £15. It took Dodgson many months of experimentation to become adept – but this he certainly became, and if he were not known for his *Alice* books he would now be mainly remembered as one of the most important photographers of the 19th century, being one of the earliest to consider photography as an art form, rather than simply as a means for recording images. In total Dodgson took around three thousand photographs, most of which are portraits of individuals or small groups. Some feature his Oxford contemporaries and give us a valuable insight into University life at the time, while others are of luminaries such as Michael Faraday, Ellen Terry, John Ruskin, and William Holman Hunt. Others are of family members, and many are of children.[18]

Several of Dodgson's earliest attempts at photography took place in the Deanery garden at Christ Church in 1856. It was here that he became acquainted with the Liddell children, Alice, Edith, Lorina, and Harry. He took many fine pictures of them, and they quickly became firm friends. They used to enjoy helping him with his photographic experiments, and he enjoyed showing them around Oxford, visiting the colleges and museums and pointing out things of interest.

Charles Dodgson polishes his camera lens

Alice, Lorina, and Edith Liddell

Boating trips on the river were a particular delight for the children, especially when Dodgson made up stories to entertain them. The most celebrated of these excursions took place on 4 July 1862 when he and his friend the Revd Robinson Duckworth, Fellow of Trinity College, took the Liddell sisters up the river to Godstow; Alice was then aged 10. It was on this occasion that Dodgson invented the story of *Alice's Adventures in Wonderland*, and Alice persuaded him to write it up:[19] the resulting manuscript, entitled 'Alice's Adventures Under Ground', was presented to Alice at Christmas-time 1864. Renamed with its usual title it was published in 1865 and has never been out of print. Its celebrated sequel, *Through the Looking-Glass and What Alice Found There*, was published in 1871.

In 1852, when Dodgson was awarded a Studentship at Christ Church, he promised to proceed to holy orders. The first stage was to become a Deacon, and in August 1861 he offered himself for examination, being ordained at a service on 22 December in Christ Church Cathedral. It was conducted by the Bishop of Oxford, Samuel Wilberforce, who is best remembered for his part in a celebrated debate with Thomas Huxley on Darwin's recent theory of evolution during a British Association meeting in Oxford.

Dodgson never proceeded to the priesthood, having come to believe that it was his duty *not* to do so, since regular parochial duties would surely take him away from the teaching career to which he felt he had been called. But there were two other reasons why he felt unable to proceed. One was his love of theatres, which were widely regarded as places of ill-repute, so when Bishop Wilberforce declared that visiting them was an absolute disqualification for parochial clergy, there was no more to be said. Another reason was that Dodgson had a speech hesitation. Although this was less serious than many have claimed, it caused him difficulties when reading in public. As a result he confined his efforts to preaching lengthy sermons, where he could make his own selection of words.[20]

In the summer of 1867 Dodgson commenced his only trip abroad – a two-month visit to the Continent with his Oxford friend the Revd Henry Liddon, who later became a Canon of St Paul's Cathedral in London. Their ambitious itinerary included overnight stops at Brussels, Cologne, Berlin, Danzig, Königsberg, St Petersburg, Moscow, Nijni Novgorod, Warsaw, Breslau, Leipzig, Giessen, Ems, and Paris.

Mathematical writings and college life

Meanwhile, Dodgson had begun to produce mathematical pamphlets to help students and books designed for the mathematically interested public. In 1861, arising from his college teaching in pure mathematics, he wrote his *Notes on the First Part of Algebra*, and followed this with pamphlets in algebra and trigonometry designed for Oxford examination

FORMULÆ.

e, as series	$1 + 1 + \frac{1}{\underline{2}} + \frac{1}{\underline{3}} + \ldots$
in decimals	$2.718281828\ldots$
e^x, as series	$1 + x + \frac{x^2}{\underline{2}} + \frac{x^3}{\underline{3}} +$
$\log_e a$, do.	$(a-1) - \frac{(a-1)^2}{2} + \frac{(a-1)^3}{3} - \ldots$
a^x, do.	$1 + \log_e a \cdot x + \frac{(\log_e a)^2 \cdot x^2}{\underline{2}} + \ldots$
$\log_e(a+1)$ do.	$a - \frac{a^2}{2} + \frac{a^3}{3} - \ldots$
$\log_e(a+1) - \log_e a$	$2\left\{ \frac{1}{2a+1} + \frac{1}{3 \cdot (2a+1)^3} + \frac{1}{5 \cdot (2a+1)^5} + \ldots \right.$
$\log_e 10$, in decimals	2.3025851
$\log_{10} e$, do	$.4342945$
$\cos\theta$, in terms of θ	$1 - \frac{\theta^2}{\underline{2}} + \frac{\theta^4}{\underline{4}} - \ldots$
$\sin\theta$, do	$\theta - \frac{\theta^3}{\underline{3}} + \frac{\theta^5}{\underline{5}}$
$\cos\theta$, exponential value	$\frac{e^{\theta i} + e^{-\theta i}}{2}$, $\left[i = \sqrt{-1} \right]$
$\sin\theta$, do.	$\frac{e^{\theta i} - e^{-\theta i}}{2i}$
$\tan^{-1}x$, in terms of x	$x - \frac{x^3}{3} + \frac{x^5}{5} - \ldots$
π, approximate values	$\frac{22}{7}, \frac{355}{113}$
in decimals	3.1415927
$\frac{\pi}{180}$, do.	$.0174533$
$\frac{180}{\pi}$, do.	57.2957795

$Mar.\ 19,\ 1878.$

A cyclostyled sheet of mathematical formulas, produced by Dodgson in 1878

[From the private collection of David and Denise Carlson]

candidates. Much of his college teaching involved geometry, on which he wrote several books and pamphlets, culminating in 1879 with his celebrated *Euclid and his Modern Rivals* (see Chapter 2).

In December 1867 Dodgson published his most important algebra book, *An Elementary Treatise on Determinants, with their Application to Simultaneous Linear Equations and Algebraical Geometry*. He had struggled with this book for almost two years, and it included some important new ideas that Bartholomew Price presented on Dodgson's behalf to the

Royal Society of London (see Chapter 3). Unfortunately, it was not a success, partly due to a lack of distribution to key mathematicians, but also because his terminology and notation were cumbersome and his over-formal approach made the book difficult to read.

On occasions Dodgson's mathematical wit extended to University issues. The early 1860s witnessed much discussion about the stipend of the controversial Regius Professor of Greek, the Revd Benjamin Jowett, which had remained unchanged at £40 since the 16th century. The matter came to a head in 1865 when substantial increases were proposed. One consequence was a humorous pseudo-mathematical pamphlet, *The New Method of Evaluation as Applied to π, an excursus to The Dynamics of a Parti-cle*, in which Dodgson presented a range of entertaining arguments to evaluate 'π', representing Jowett's salary.[21]

In June 1868 Charles's father, Archdeacon Dodgson, died suddenly at Croft Rectory. As the head of the large family, Charles was now responsible for finding a new home for his many brothers and sisters. After a short search they leased The Chestnuts, a substantial house in Guildford, in Surrey. The family left Croft in September and by Christmas were happily settled in their new surroundings. But family life was not the only change for Dodgson at that time. Although his teaching continued much as before, there were other changes in college life and in the direction of his studies.

In October 1868 Dodgson moved into a new suite of rooms in college. Up to this time priority for certain rooms had been assigned to the noblemen who resided in college, but a change of rules resulted in a magnificent suite becoming available. Dodgson thereby acquired the finest and most expensive accommodation in the college, in the north-west corner of Tom Quad; these included a living room, a dining room, and a bedroom. With the college's permission, he constructed a fine photographic studio on the roof above his rooms. He also purchased some tiles by the artist William De Morgan, son of the Victorian mathematician Augustus De Morgan, for the fireplace in his living room.

Dodgson's living room
at Christ Church

[Governing Body of Christ
Church, Oxford]

Dodgson was now playing an increasing part in the life of the college. In October 1867, after much discussion and the passage of The Christ Church Ordinances (Oxford) Bill through Parliament, the college had a new Governing Body – previously all decisions had been made by the Dean and Canons. He thus became involved with such matters as the selection of Senior Students and the appointment of Fellows, as well as with decisions on the college buildings.

In the 1870s, mainly through such college events as the selection of an architect for a new belfry, Dodgson increasingly became involved in elections, ranking procedures, and the theory of voting. These topics had been studied extensively by the Marquis de Condorcet around the time of the French Revolution, and Dodgson's original contributions to them, among the most creative of all his mathematical investigations, have been described as the most important after Condorcet (see Chapter 5).

In 1880 the college suffered a financial crisis. At the beginning of February Dodgson wrote to the college authorities proposing that his salary should be lowered from £300 per year to £200. A year later he decided to give up the Mathematical Lectureship he had held for twenty-five years in order to devote more time to the books and articles that already brought him a sufficient income. On 18 October 1881 he resigned in order to do more writing,[22]

partly in the cause of Mathematical education, partly in the cause of innocent recreations for children, and partly, I hope (though so utterly unworthy of being allowed to take up such work) in the cause of religious thought.

His last official lecture was on 30 November 1881, when he recalled:[23]

I gave my first Euclid Lecture in the Lecture-room on Monday January 28, 1856. It consisted of 12 men, of whom nine attended.

Final years

In late 1882 the Curator of the Christ Church Common Room resigned, having held the position for twenty-one years, and Dodgson reluctantly agreed to take on the job. It was a time-consuming burden, involving the day-to-day management of the Common Room. Dodgson became Curator for nine years.

Voting theory

The period after the 1880 General Election was an interesting time in British politics. There were two main parties: the Liberals, who had been returned to power with William Gladstone as Prime Minister, and the Conservatives, with Lord Salisbury taking over as leader following the death of Benjamin Disraeli in 1881. Much had changed over the previous fifty years, with the Reform Acts of 1832 and 1867 extending the franchise to merchants and industrial workers, and the Third Reform Bill of 1884 extended it still further to agricultural workers and miners. Fewer than 200 of the 658 Members of Parliament elected in 1880 represented single-member electoral districts, while over 400 were in two-member constituencies and a handful of districts returned three or four members. The Third Reform Bill proposed a massive increase in the number of single-member districts, and at the 1885 General Election 616 members became the sole representatives of their constituents.

In his desire to achieve a fairer representation of views, Dodgson was bitterly opposed to these single-member districts. In electoral districts with members from different parties, both majority and minority views could be represented, but in a single-member district with roughly equal numbers of supporters from each side, the views of almost half of the electorate were unrepresented. Dodgson's own preference was for electoral districts with two to five members, in which electors were given just one vote.

In 1884 the Proportional Representation Society, seeking the best way of allocating seats in multi-member constituencies, proposed reform on the basis of a single transferable vote (see Chapter 5). Dodgson had other ideas, and in November of that year he published *The Principles of Parliamentary Representation*, his major work on the subject. This pamphlet was later described enthusiastically by the political economist Duncan Black as 'the most interesting contribution to Political Science that has ever been made'.[24]

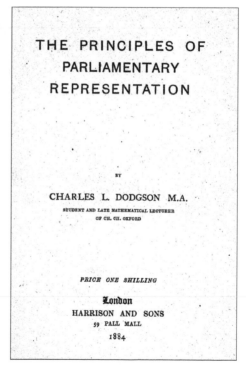

Some of Dodgson's political recommendations were eventually adopted in Britain, such as the rule that no results can be announced until all the voting booths have closed. Others, such as his proposals for proportional representation, were not. In the 1870s Dodgson had announced his intention to write a book on voting and elections, but such a publication never materialized, causing the Oxford philosopher Michael Dummett to remark:[25]

Dodgson's 1884 *The Principles of Parliamentary Representation*

It is a matter of the deepest regret that Dodgson never completed the book that he planned to write on the subject. Such were his lucidity of exposition and his mastery of the topic that it seems possible that, had he ever published it, the political history of Britain would have been significantly different.

Logic

Dodgson's interest in logic dated from his undergraduate days when he was required to sit a logic paper as part of his Classical examinations. The subject permeated all aspects of his life, and references to logic in Dodgson's diaries occur as early as 6 September 1855, where he records that he 'Wrote part of a treatise on Logic, for the benefit of Margaret and Annie Wilcox'.[26] Later, from around 1885 until his untimely death in 1898, he spent much of his time presenting symbolic logic as an entertainment for children to develop their powers of logical thought, and as a serious topic of study for adults. In particular, he was so keen to introduce young people to the delights of symbolic logic that he presented classes on logic at the Oxford High School for Girls and in two of the Oxford colleges. In order to increase their circulation, he published some of his writings on logic under his pen-name of Lewis Carroll.

For some time Dodgson had nursed the idea of writing a multi-volume work on symbolic logic (or 'Logic for Ladies' as he originally intended to call it), to begin with a small pamphlet called *The Game of Logic*, designed to convey the ideas of syllogisms to young people. By the end of the year this small pamphlet had expanded to a book of about one hundred pages. More serious was *Symbolic Logic. Part I. Elementary*, which appeared in February 1896, and the five hundred copies sold out immediately. Promoted as 'A fascinating mental recreation for the young', the work was 'Dedicated to the memory of Aristotle' and opens with some encouragement for his young readers:[27]

If, dear Reader, you will faithfully observe these Rules, and so give my book a really fair trial, I promise you, most confidently, that you will find Symbolic Logic to be one of the most, if not *the* most, fascinating of mental recreations!…I have myself taught most of its contents, *vivâ voce*, to *many* children, and have found them take a real intelligent interest in the subject.

Sadly, Dodgson died before Part II of his *Symbolic Logic* was completed. As we see in Chapter 4, this included examples with as many as fifty propositions, a general method he devised for solving puzzles of this kind, and a number of logical puzzles and paradoxes. Although parts of his manuscript were in galley proof form, these disappeared and were not rediscovered until many years after his death.

A CHALLENGE TO LOGICIANS.

GIVEN

1. If some *a* are *b* and some not, some *c* are not *d*;

2. If some *e* are *f*, and if some *g* are *h*, some *j* are *k*;

3. If all *l* are *m*, no *n* are *p*;

4. If some *c* are *d* and some not, some *g* are *h*;

5. If no *e* are *f*, and if some *n* are *p*, some *j* are not *k*;

6. If some *e* are not *f*, and if some *g* are not *h*, some *n* are *p*;

7. If some *c* are not *d*, and if some *j* are *k*, no *e* are *f*;

8. If some *g* are not *h*, and if some *j* are not *k*, some *l* are *m*;

9. If some *e* are not *f*, and if some *n* are *p*, some *a* are not *b*;

10. If some *a* are *b*, and if some *c* are *d*, some *g* are not *h*;

11. If some *c* are not *d*, and if some *l* are not *m*, some *e* are *f*:

TO PROVE

If some *a* are *b*, and if some *e* are not *f*, no *c* are *d*.

C. L. Dodgson.

Ch. Ch., Oxford,
Oct. 1892.

[N. B. Copies of this paper may be had by applying to
Messrs. Parker, Broad Street.]

A logical puzzle, posed by Dodgson in 1892
[Governing Body of Christ Church, Oxford]

Recreational mathematics

But symbolic logic was not Charles Dodgson's only preoccupation in the last few years of his life. In the early 1880s, under his pen-name of Lewis Carroll, he wrote a puzzle column for a periodical called *The Monthly Packet*, with each issue featuring a story that concealed some ingenious mathematical problems. The ten stories, called 'Knots', were subsequently collected together into a puzzle book, *A Tangled Tale*, which appeared in time for Christmas 1885 (see Chapter 6).

Another activity was to produce a substantial collection of mathematical problems, *Curiosa Mathematica Part II. Pillow-Problems Thought Out During Sleepless Nights*, which appeared in 1893. In the Introduction the author described how it came into being:[28]

Nearly all of the following seventy-two Problems are veritable "Pillow-Problems", having been solved, in the head, while lying awake at night…every one of them was worked out, to the very end, before drawing any diagram or writing down a single word of the solution. I generally wrote down the *answer*, first of all: and *afterwards* the question and its solution.

In December 1897 Dodgson travelled to Guildford to spend Christmas at The Chestnuts with his sisters. His brief diary entry for 23 December 1897 was the last that he wrote: 'I start for Guildford by the 2.07 today'.[29] While in Guildford he worked hard on Part II of his *Symbolic Logic*, hoping to finish it. But on 6 January he developed a feverish cold that rapidly developed into severe bronchial influenza. He died on 14 January 1898 at the age of 65, and was buried in The Mount Cemetery in Guildford. In the same week Henry Liddell, whose daughter Alice had been so much a part of Dodgson's life, also died, and a joint memorial service was held for them at Christ Church Cathedral on Sunday 23 January. As Dean Paget recalled:[30]

Two memorial windows in All Saints' Church, Daresbury, Cheshire

Within the last ten days Christ Church has lost much. And though the work that bore the fame of Lewis Carroll far and wide stands in distant contrast with the Dean's, still it has no rival in its own wonderful and happy sphere; and in a world where many of us laugh too seldom, and many of us laugh amiss, we all owe much to one whose brilliant and incalculable humour found us fresh springs of clear and wholesome and unfailing laughter.

Dodgson, mathematics, and mathematicians

As we have seen, mathematics was a lifelong interest of Charles Dodgson. Although he explored numerous topics, a constant feature of his interest in the subject seems motivated by its absolute certainty and its suitability for training the mind. In *A New Theory of Parallels*, Dodgson contrasted the certainty of mathematics with the situation in the natural sciences:[31]

It may well be doubted whether, in all the range of Science, there is any field so fascinating to the explorer − so rich in hidden treasures − so fruitful in delightful surprises − as that of Pure Mathematics. The charm lies chiefly, I think, in the absolute *certainty* of its results: for that is what, beyond almost all mental treasures, the human intellect craves for…Most other Sciences are in a state of constant flux − the precious truths of one generation being smiled at as paradoxes by the second generation, and contemptuously swept away as childish nonsense by the third. If you would see a specimen of the rapidity of this process of decomposition, take Biology for a sample: quote, to any distinguished Biologist you happen to meet, some book published thirty years ago, and observe his pitying smile!

But neither thirty years, nor thirty centuries, affect the clearness, or the charm, of Geometrical truths. Such a theorem as 'the square of the hypotenuse of a right-angled triangle is equal to the sum of the squares of the sides' is as dazzlingly beautiful now as it was in the day when Pythagoras first discovered it, and celebrated its advent, it is said, by sacrificing a hecatomb of oxen [100 oxen] − a method of doing honour to science that has always seemed to me slightly exaggerated and uncalled-for.

There has been dispute among historians about Dodgson's mathematical standing. Dodgson himself did not make high claims about his own mathematical work, considering himself primarily as a mathematical teacher who investigated some specific topics such as the theory of determinants, Euclidean geometry, and symbolic logic. Within these areas, in which he believed to have some expertise, he did not hesitate to challenge other mathematicians whom he considered as much more distinguished than he was.

A powerful illustration is given by Dodgson's criticism of James Maurice Wilson, a mathematician who played a significant role in the revolt against Euclid (see Chapter 2). In his introduction to his *Euclid and his Modern Rivals*, Dodgson apologized to Wilson:[32]

To Mr. Wilson especially such apology is due…partly because it may well be deemed an impertinence in one, whose line of study has been chiefly in the lower branches of Mathematics, to dare to pronounce any opinion at all on the work of a Senior Wrangler. Nor should I thus dare, if it entailed my following him up 'yonder mountain height' which *he* has scaled, but which *I* can only gaze at from a distance: it is only when he ceases 'to move so near the heavens,' and comes down into the lower regions of Elementary Geometry, which I have been teaching for nearly five-and-twenty years, that I feel sufficiently familiar with the matter in hand to venture to speak.

Yet after examining Wilson's manual, offered as a substitute to Euclid's *Elements*, Dodgson made a harsh assessment:[33]

ELEMENTARY GEOMETRY

PART I.

ANGLES, PARALLELS, TRIANGLES, EQUIVALENT FIGURES, WITH THE APPLICATION TO PROBLEMS.

COMPILED BY

J. M. WILSON, M.A.
FELLOW OF ST JOHN'S COLLEGE, CAMBRIDGE,
AND MATHEMATICAL MASTER OF RUGBY SCHOOL.

London and Cambridge:
MACMILLAN AND CO.
1868

[*All Rights reserved.*]

J. M. Wilson and his 'rival text' *Elementary Geometry*

The abundant specimens of logical inaccuracy, and of loose writing generally, which I have here collected would, I feel sure, in a mere popular treatise be discreditable – in a scientific treatise, however modestly put forth, deplorable – but in a treatise avowedly put forth as a model of logical precision, and *intended to supersede Euclid*, they are simply monstrous.

My ultimate conclusion on your Manual is that it has *no claim whatever* to be adopted as *the* Manual for purposes of teaching and examination.

Dodgson was first and foremost a mathematical teacher, and it is as such that his work should be read and his mathematical activity understood. As Adrian Rice and Eve Torrance have noted:[34]

Dodgson was certainly not what could be described as an active research mathematician. Indeed, he did not belong to any mathematical or scientific societies, nor did he subscribe to the major mathematics research journals of the day.

But although Dodgson joined none of the mathematical societies of his time, we should note that there were no societies specifically for mathematics education until the establishment in 1871 of The Association for the Improvement of Geometrical Teaching (see Chapter 2).[35] This association was founded by teachers who challenged the use of Euclid's *Elements* as a manual for geometrical teaching, and it is not surprising that Dodgson did not join it, being a fierce supporter of Euclid. Indeed, he was strongly opposed to Euclid's rivals and renamed their society the 'Association for the Improvement of Things in General'.[36]

The idea of a mathematical society of Euclid's supporters was advanced by Charles Taylor in 1873, during the British Association meeting in Bradford, but the project was not achieved.[37] We may imagine that Dodgson would have been interested in such an association, and he certainly was not opposed to the idea of associations that suited him. As his bank account shows, he financially supported numerous societies: these included charitable ones (The Strangers' Friend Society), social ones (The Associate Institution for Improving and Enforcing the Laws for the Protection of Women), religious ones (The British & Foreign Bible Society), and literary ones (The Pure Literature Society).[38] Dodgson also joined the Society for Psychical Research at its foundation in 1882 and remained a member until his death in 1898.[39] The president of that society at its foundation was the Cambridge philosopher Henry Sidgwick, with whom Dodgson corresponded on logic, and the Society counted among its members such celebrities as Arthur Conan Doyle, Mark Twain, Arthur Balfour, William Crookes, and John Venn.[40]

There is no evidence that Dodgson subscribed to, or regularly read, the main mathematical journals of his time. However, the sale catalogues of his library show that he owned sets of journals that circulated among mathematical teachers and junior mathematicians and also regularly attracted confirmed mathematicians. For example, Dodgson owned at least sixty-seven volumes of *Mathematical Questions and Solutions*, the mathematical supplement to *The Educational Times*, which devoted its columns to mathematical problems sent in and solved by its readers.[41] This journal attracted a large readership among mathematical teachers and students in Britain and abroad, and received contributions from some of the leading mathematicians of the time, including Arthur Cayley, William K. Clifford, James J. Sylvester, and Bertrand Russell,[42] and Dodgson himself made several contributions to it. He also owned copies of journals that were mainly addressed to junior mathematicians, such as the *Messenger of Mathematics* (the joint mathematical journal of the Universities of Oxford, Cambridge, and Dublin) and the *Quarterly Journal of Pure and Applied Mathematics* (a revival of the *Cambridge and Dublin Mathematical Journal*).[43]

An examination of Dodgson's private library also gives an idea of his mathematical readings. It has long been claimed that he worked alone and read little of the work of his contemporaries, as Derek Hudson observed in his biography of Dodgson:[44]

> He read comparatively little of the works of other mathematicians or logicians, preferring to develop his theories out of his own mind. This method had its advantages, no doubt, yet it not only gave him a lot of unnecessary trouble but deprived him of the chance of escaping avoidable mistakes. In fact he handled scientific matters in the same way as he dealt with controversial language, and the method was never likely to produce – nor did it produce – a mathematical achievement of comparable value, in its own line, to *Alice in Wonderland*.

However, the references cited in his writings show that he knew many of the main British mathematical authors of his time, such as George Boole, Augustus De Morgan, George Peacock, George Salmon, Henry J. S. Smith, Isaac Todhunter, and William Spottiswoode, and even though Cayley, Sylvester, and Clifford are not found among the sale

MATHEMATICAL

QUESTIONS AND SOLUTIONS,

FROM THE "EDUCATIONAL TIMES,"

WITH MANY ADDITIONAL

PAPERS AND SOLUTIONS

NOT PUBLISHED IN THE "EDUCATIONAL TIMES,"
AND

AN APPENDIX.

EDITED BY

W. J. C. MILLER, B.A.,

REGISTRAR
OF THE
GENERAL MEDICAL COUNCIL.

VOL. XLIII.

LONDON:
FRANCIS HODGSON, 89 FARRINGDON STREET, E.C.

1885.

Mathematical Questions and Solutions, from the "Educational Times"

Edward Sampson and Robert Baynes
[Senior Common Room of Christ Church, Oxford]

Henry Smith

catalogues of his library it is likely that Dodgson knew of them. He met Cayley in Oxford, an instance that he recorded in his diary on 7 June 1864,[45] while Sylvester is mentioned in his exchanges with Spottiswoode on determinants (see Chapter 3).[46]

It remains true that Dodgson was outside the main mathematical circles and could not claim familiarity with the major mathematicians of his time (Smith possibly being excepted), and yet one should not infer that he worked alone and isolated. Indeed, his diaries, letters, and papers show that he regularly solicited the opinions of Oxford colleagues concerning mathematical topics in which he was interested. This close Oxonian circle included Bartholomew Price, Edward F. Sampson, Robert E. Baynes, occasionally Henry Smith, and later John Cook Wilson on logic. He also sometimes sought the opinions of his Cambridge colleagues, such as Todhunter, Sidgwick, and Venn.

Even though Dodgson seems not to have been a regular reader of the general mathematical literature of his time, he was certainly familiar with published writings in his specific areas of interest. This is evident in the case of geometry and his defence of Euclid: indeed, he gathered dozens of books that were offered as 'rivals' to Euclid, both in Britain and abroad, and refuted the main ones in his *Euclid and his Modern Rivals*. His late interest in symbolic logic can also be seen through his acquisition of the main books on the topic that were published in his lifetime, even though it is not clear how familiar he was with them (see Chapters 4 and 7).[47]

Dodgson's mathematical standing cannot be understood without a proper consideration of the types of mathematical problem that he investigated and the motivations that made him do so. The following chapters address five major areas to which he contributed: geometry, algebra, logic, the theory of voting, and recreational mathematics. We conclude with a discussion of his mathematical legacy and a bibliography of his mathematical writings.

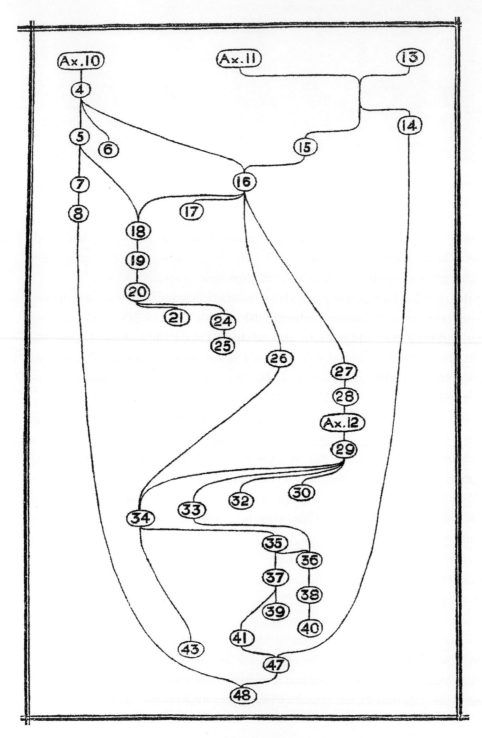

Dodgson's hierarchical arrangement of the propositions in Euclid's *Elements*, Book I, from the frontispiece to *Euclid, Books I, II*

CHAPTER 2

Geometry

ROBIN WILSON

C harles Dodgson was fascinated by geometry. As we saw in Chapter 1, the 12-year-old's remarkable two-page document on trisecting a right angle illustrated how at an early age he had become conversant with the fundamentals of the subject, as exhibited in such classical works as Euclid's *Elements*, the most important geometry book of all time.

Throughout his life Dodgson was involved with the *Elements*. As a pupil at Rugby School he was required to learn propositions from it, and as an Oxford lecturer he taught its contents to undergraduates preparing for their examinations. Many of his publications were devoted to explaining it, championing it when it was under threat, and even referring to it humorously, as in his parody 'Hiawatha's Photographing'[1] where he described Hiawatha's rosewood camera as looking

…all squares and oblongs,
Like a complicated figure
In the second book of Euclid.

But who was Euclid, and why was his *Elements* so important to Dodgson?

Euclid's *Elements*

Euclid wrote his *Elements* in Alexandria (then in the Greek world, but now in Egypt) during the 3rd century BC. For over 2000 years – from the ancient Greek academies, the

The Mathematical World of Charles L. Dodgson (Lewis Carroll). Robin Wilson and Amirouche Moktefi.
Oxford University Press (2019). © Oxford University Press 2019.
DOI: 10.1093/oso/9780198817000.001.0001

Islamic world, and the universities of medieval Europe to the private schools of Victorian England – versions of this classic text were used to teach geometry and train the mind. Indeed, as we shall see, the *Elements* was one of the most printed works of all time, possibly coming only second to the Bible: during the 19th century over two hundred editions were published in England alone, with one popular version selling over half a million copies.

Euclid's *Elements* consists of thirteen 'Books':[2]

The thirteen books of Euclid's *Elements*

Book I: Foundations of plane geometry
Book II: The geometry of rectangles
Book III: The geometry of the circle
Book IV: Regular polygons in circles
Book V: The general theory of magnitudes in proportion
Book VI: The plane geometry of similar figures
Book VII: Basic arithmetic
Book VIII: Numbers in continued proportion
Book IX: Odd and even numbers and prime and perfect numbers
Book X: Incommensurable line segments
Book XI: Foundations of solid geometry
Book XII: Areas and volumes; Eudoxus' method of exhaustion
Book XIII: The Platonic solids

Books I and II introduce the geometry of the plane – points, lines, angles, parallel lines, triangles, rectangles, and parallelograms. A typical result on triangles is Book I, Proposition 5, the *pons asinorum* or asses' bridge on isosceles triangles (those with two sides of equal length), which states that

In isosceles triangles the angles opposite the equal sides are equal to one another.

In medieval European universities, if you understood this proposition then you could cross the asses' bridge and proceed to the delightful results beyond, such as the *Pythagorean theorem* on the areas of squares on the sides of a right-angled triangle (Book I, Proposition 47):

In a right-angled triangle the square on the side subtending the right angle is equal to the sum of the squares on the sides containing the right angle.

Here, area Z = area X + area Y.

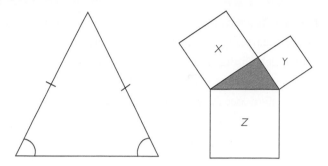

The *pons asinorum* and the Pythagorean theorem

Books III and IV are concerned with circles in the plane. A typical result from Book III, Proposition 31, concerns a triangle drawn inside a semicircle with one side as a diameter:

In a circle the angle in a semicircle is a right angle.

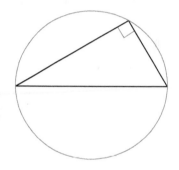

Books V and VI are concerned with magnitudes in proportion and with similar figures – those of the same shape but not necessarily the same size. In Greek geometry, numbers were usually thought of, and represented by, lengths of finite line segments, and a typical deduction in Book V is Proposition 16:

If four magnitudes be proportional, they will also be proportional alternately.

Expressed in terms of ratios of lengths, this states that:

If the ratio $a : b$ equals the ratio $c : d$, then the ratio $a : c$ equals the ratio $b : d$.

For example, knowing that $6 : 8 = 9 : 12$, we can deduce that $6 : 9 = 8 : 12$.

These books also discuss 'commensurable' magnitudes – pairs of lengths whose ratio can be written in terms of whole numbers, such as $0.05 : 0.04$ or $5\pi : 4\pi$, with ratio $5 : 4$. Two incommensurable lengths are the side and diagonal of a square, because these lengths are in the ratio of $1 : \sqrt{2}$, which cannot be written in terms of whole numbers.

Books VII, VIII, and IX deal with the arithmetic of odd and even numbers, squares and cubes, and prime and perfect numbers. In particular, Book VII presents the Euclidean algorithm for finding the greatest common divisor of two numbers, and Book IX contains Euclid's celebrated proof that there are infinitely many primes.

Book X is a long and complicated account of commensurable numbers.

Books XI, XII, and XIII present the geometry of three dimensions – solids, pyramids, spheres, cones, and cylinders. In particular, Book XIII introduces the *Platonic solids*, the regular polyhedra in which all the faces and all the corners look the same, and includes a proof that there can be only five of these: the tetrahedron (four faces), the cube (six faces), the octahedron (eight faces), the dodecahedron (twelve faces), and the icosahedron (twenty faces).

The structure of the *Elements*

Euclid's *Elements* is organized in a hierarchical way, building on certain initial statements called *definitions*, *postulates*, and *common notions* (Euclid's term for axioms). Here are two typical definitions from the beginning of Book I:[3]

8. A *plane angle* is the inclination of two straight lines to one another, which meet together, but which are not in the same direction.
15. A *circle* is a plane figure contained by one line such that all the straight lines falling upon it from one point among those lying within the figure are equal to one another.

Dodgson described a postulate as 'something to be done, for which no proof is given'.[4] Euclid listed five postulates, the first three of which enable us to carry out 'ruler and compass' constructions that permit the use of only an unmarked ruler for drawing straight lines and a pair of compasses for drawing circles:[5]

    Let the following be postulated:

1. To draw a straight line from any point to any point.
2. To produce a finite straight line continuously in a straight line.
3. To describe a circle with any centre and distance.

The fourth postulate states that all right angles are to be considered equal to one another, while the fifth postulate was considerably different in character and we return to it later.

    Dodgson also described an axiom as 'something to be believed, for which no proof is given'.[6] Euclid listed five axioms, starting with the following:

1. Things which are equal to the same thing are also equal to each other.
2. If equals be added to equals, the wholes are equal.

In modern notation, these say:

1. If $a = x$ and $b = x$, then $a = b$,
2. If $a + b = x$ and $c + d = y$, then $a + b + c + d = x + y$.

From these humble beginnings Euclid first deduced some very simple results, then more complicated ones, and so on, until he had created the enormous hierarchical structure that we now call *Euclidean geometry*; the figure at the beginning of this chapter shows Dodgson's hierarchical arrangement of the forty-eight propositions in Book I. Most of Euclid's propositions were theorems to be proved, while others involved ruler-and-compass constructions whose validity was then justified. An example of the latter is Book I, Proposition 1, with appropriate references to preceding definitions, postulates, and the first axiom.

Euclid's *Elements*, Book I, Proposition I

On a given finite straight line to construct an equilateral triangle.

Let *AB* be the given finite straight line.

Thus it is required to construct an equilateral triangle on the straight line *AB*.

With centre *A* and distance *AB* let the circle *BCD* be described;   [Post. 3]

again, with centre *B* and distance *BA* let the circle *ACE* be described;

and from the point *C*, in which the circles cut one another, to the points *A*, *B* let the straight lines *CA*, *CB* be joined.   [Post. 1]

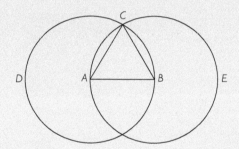

Now, since the point *A* is the centre of the circle *CDB*, *AC* is equal to *AB*.   [Def. 15]

Again, since the point *B* is the centre of the circle *CAE*, *BC* is equal to *BA*.   [Def. 15]

But *CA* was proved equal to *AB*; therefore each of the straight lines *CA*, *CB* is equal to *AB*.

And things which are equal to the same thing are also equal to one another.   [C.N. 1]

Therefore *CA* is also equal to *CB*.

Therefore the three straight lines *CA*, *AB*, *BC* are equal to one another.

Therefore the triangle *ABC* is equilateral; and it has been constructed on the given finite straight line *AB*.

Being what it was required to do.[7]

Editions of Euclid's *Elements*

No original copies of Euclid's *Elements* survive, and for early versions we have to rely on Greek and Islamic commentaries and on Arabic, Latin, and Greek translations. The oldest extant copy is a Greek text, copied in Constantinople in the year 888, which is housed in the Bodleian Library in Oxford.

The first printed edition of Euclid's *Elements* was a Latin edition, published in Venice in 1482, and the first English version was Henry Billingsley's translation of 1570, with a preface by John Dee. Following this, there were hundreds of printed editions in many languages, differing greatly in style and content and often containing only the first six Books. A celebrated example is Oliver Byrne's 1847 edition of Euclid's *Elements*, Books I–VI, in which many of the geometrical terms were replaced by small coloured diagrams.

In 1756 Robert Simson, who had taught at the University of Glasgow, produced a version of Euclid's *Elements* that was used for many years, achieving its 26th edition in 1844. As Francine Abeles has described it:[8]

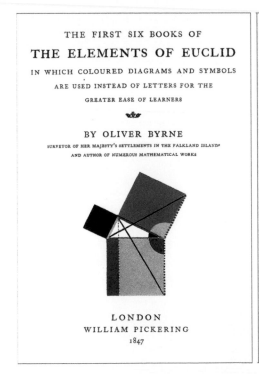

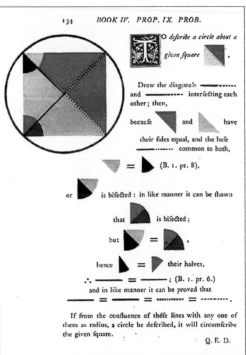

Two pages from Byrne's colour edition of Euclid's *Elements*

Simson captured the spirit of Euclid, and his book was the standard geometry textbook of England. His version of Euclid was excellent because it was based on an authoritative Latin translation, not because Simson himself was a superb textual critic of Euclid…He was considered an excellent interpreter as well as critic of Euclid.

For his edition, Simson renumbered some of Euclid's definitions, axioms, and postulates; for example, postulates 4 and 5 became axioms 11 and 12. Many of the two hundred or more editions of Euclid's *Elements* that were published in England in the 19th century were based on this work and used his numbering.

Over the Easter vacation of 1855 Charles Dodgson started teaching geometry to Louisa, the most mathematically gifted of his sisters, commenting on one of these later editions:[9]

Went into Darlington – bought at Swale's, *Chamber's Euclid* for Louisa. I had to scratch out a good deal he had interpolated, (e.g. definitions of words of his own) and put some he had left out. An author has no right to *mangle* the original writer whom he employs: all additional matter should be carefully distinguished from the genuine text. N.B. Pott's *Euclid* is the only edition worth getting – both Capell and Chamber's are mangled editions.

Here Dodgson's preferred choice of 'Pott's *Euclid*' was Robert Potts's edition, *Euclid's Elements of Geometry*, based on the text of Robert Simson and first published in 1845.[10] An 1861 guide to the Honours courses in Oxford commented on its use in the teaching of geometry:[11]

Geometry must, of course, be commenced by acquiring a thorough knowledge of Euclid, which will be best read in Mr. Pott's octavo edition. The VIth book should not be read until the principles of ratio and proportion enunciated in the Vth are entirely mastered. It is not usual to read more than the first twenty propositions of the XIth book; and the XIIth may be said to be completely superseded by other methods. In Mr. Pott's Appendix is a short but useful chapter on transversals.

Another popular edition was published by Isaac Todhunter, a Cambridge Senior Wrangler, Fellow of the Royal Society, and prolific writer of textbooks for schools and colleges. His text, *The Elements of Euclid*, also based on Simson's text, was first published in 1862 and sold half a million copies.

Were there too many editions of the *Elements*? One person who thought so was Hugh MacColl (see Chapter 7), a mathematical reviewer for the *Athenaeum*, who was underwhelmed by an 1890 edition by A. E. Layng, Headmaster of Stafford School:[12]

Why so many Euclids and all running in the old grooves? Mr. Layng's is carefully compiled and well arranged, but since the same praise is due to so many others, why add to the interminable list?

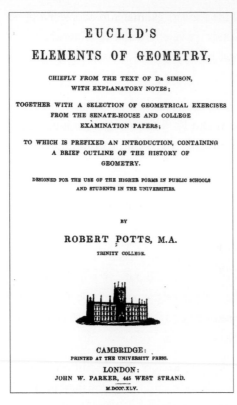

EUCLID'S
ELEMENTS OF GEOMETRY,

CHIEFLY FROM THE TEXT OF DR SIMSON,
WITH EXPLANATORY NOTES;

TOGETHER WITH A SELECTION OF GEOMETRICAL EXERCISES
FROM THE SENATE-HOUSE AND COLLEGE
EXAMINATION PAPERS;

TO WHICH IS PREFIXED AN INTRODUCTION, CONTAINING
A BRIEF OUTLINE OF THE HISTORY OF
GEOMETRY.

DESIGNED FOR THE USE OF THE HIGHER FORMS IN PUBLIC SCHOOLS
AND STUDENTS IN THE UNIVERSITIES.

BY

ROBERT POTTS, M.A.

TRINITY COLLEGE.

CAMBRIDGE:
PRINTED AT THE UNIVERSITY PRESS.
LONDON:
JOHN W. PARKER, 445 WEST STRAND.
M.DCCC.XLV.

THE ELEMENTS OF

EUCLID

FOR THE USE OF SCHOOLS AND COLLEGES;

COMPRISING THE FIRST SIX BOOKS AND PORTIONS OF THE ELEVENTH
AND TWELFTH BOOKS;

WITH NOTES, AN APPENDIX, AND EXERCISES

BY

I. TODHUNTER M.A., F.R.S.

NEW EDITION.

MACMILLAN AND CO.
London and Cambridge.
1867

The Right of Translation and Reproduction is reserved.

Robert Potts's edition of *Euclid's Elements of Geometry*

Isaac Todhunter's *The Elements of Euclid*

Dodgson's geometry teaching

As we saw in Chapter 1, the Oxford University course required students to sit three main examinations – *Responsions*, *Moderations*, and *Finals*. Those offering the geometry paper in Responsions (or 'Little-Go') were expected to know the material in the *Elements*, Books I and II, passmen taking the Euclid paper in Moderations would study up to Book III, and classmen working for the Final Honours School in Mathematics would continue up to Book VI.

Much of Dodgson's teaching to Christ Church undergraduates involved teaching Euclid's *Elements* for all three examinations. His diary for 5 March 1855 includes the following entry:[13]

Got a note from Leighton, a gentleman commoner, who wishes to be taught some arithmetic for his Little-Go, (which comes in about a fortnight) – as well as the second book of Euclid. We settled that he is to come an hour an evening, and began at once: so far as we went, he is well up, and needs no teaching: he makes my third pupil.

Two weeks later Baldwin Leighton passed his Responsions exams, and on receiving his £5 payment Dodgson recorded that this was the first money he had ever earned.

Opinions differ widely as to how effective a teacher Dodgson was. One former under-graduate was complimentary:[14]

I was up at Christ Church as an undergraduate early in the eighties, he being my mathematical tutor and certainly his methods of explaining the elements of Euclid gave me the impression of being extremely lucid, so that the least intelligent of us could grasp at any rate 'the Pons Asinorum'.

But another recollection was more cryptic:

…glancing at a problem in Euclid which I had written out, he placed his finger on an omission. "I deny you the right to assert that." I supplied what was wanting. "Why did you not say so before? What is a corollary?" Silence. "Do you ever play billiards?" "Sometimes." "If you attempted a can-non, missed, and holed your own and the red ball, what would you call it?" "A fluke." "Exactly. A corollary is a fluke in Euclid. Good morning."

It seems as though the liveliness and effectiveness of his college teaching varied from pupil to pupil and from topic to topic.

Dodgson's writings on geometry

Although Dodgson greatly admired Euclid's *Elements*, he recognized that it contained gaps, inaccuracies, and inconsistencies. To help his pupils overcome these, he produced a number of books and short mathematical pamphlets that clarified the text, suggested alternative approaches, and included exercises for his pupils to try. Over the years he became an enthusiastic producer of pamphlets, publishing more than two hundred of them on a wide range of topics.

Dodgson's first mathematical pamphlet was *Notes on the First Two Books of Euclid* (1860). Costing sixpence, and written to help those taking Responsions, this was designed to be used in conjunction with one of Simson's editions. It was subsequently expanded into book form as *Euclid, Books I, II*, which appeared in an unpublished private version in 1875 and was published in 1882, eventually running to eight editions. Dodgson's aim, as explained in the Preface, was

to show what Euclid's method really is in itself, when stripped of all accidental verbiage and repetition. With this object, I have held myself free to alter and abridge the language wherever it seemed desirable, so long as I made no real change in his methods of proof, or in his logical sequence.

Dodgson's geometry books and pamphlets

1860:	*A Syllabus of Plane Algebraical Geometry*
1860:	*Notes on the First Two Books of Euclid*
1863:	*The Enunciations of Euclid I, II*
1865:	*The Dynamics of a Parti-cle*
1868:	*The Fifth Book of Euclid*
1872:	*Symbols, &c., to be Used in Euclid, Books I, II*
1872:	*Number of Propositions in Euclid*
1873:	*Enunciations of Euclid I–VI*
1874:	*Euclid, Book V*
1875/82:	*Euclid, Books I, II*
1879:	*Euclid and his Modern Rivals*
1885:	*Supplement to Euclid and his Modern Rivals*
1888:	*Curiosa Mathematica, Part I. A New Theory of Parallels*

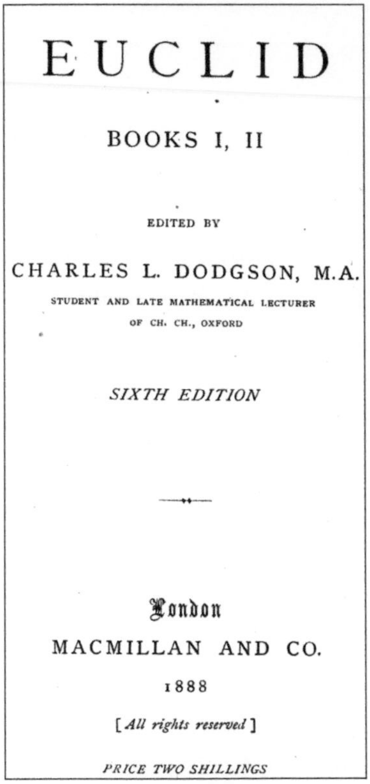

EUCLID

BOOKS I, II

EDITED BY

CHARLES L. DODGSON, M.A.

STUDENT AND LATE MATHEMATICAL LECTURER OF CH. CH., OXFORD

SIXTH EDITION

London

MACMILLAN AND CO.

1888

[*All rights reserved*]

PRICE TWO SHILLINGS

Title page of *Euclid, Books I, II*, sixth edition

The result is that the text of this Edition is (as I have ascertained by counting the words) *less than five-sevenths* of that contained in the ordinary Editions.

Associated with these Notes was a useful pamphlet for those preparing for examinations. Entitled *Enunciations of Euclid I, II* (1863), it listed all the main definitions and propositions needed for Responsions. Ten years later Dodgson expanded this to *Enunciations of Euclid I–VI* for those preparing for Moderations and Finals.

Dodgson always preferred to teach Euclid's Book V on ratio and proportion from an algebraic point of view, and in 1868 he produced a pamphlet for the Pass Schools, with complete title *The Fifth Book of Euclid Treated Algebraically, so Far as It Relates to Commensurable Magnitudes, with Notes*. In this pamphlet he explained each definition with examples and recast each proposition into a more accessible algebraic form. For simplicity he omitted the study of incommensurable numbers (such as $\sqrt{2}$) as being inappropriate for undergraduates:[15]

My reasons for this omission are two: first, that I believe it to be much too abstruse a subject for the ordinary Pass Examination; secondly, that it is not required in it. The exemption is a most necessary one, though the effect of it is to reduce the Vth and VIth Books, in the form in which they are now learned and accepted in the Schools, to a logical absurdity.

While these pamphlets and books were generally well received and widely used, not everyone was enthusiastic. In particular, Robert Potts, editor of Dodgson's preferred version of the *Elements*, had concerns about spoon-feeding the readers, as he observed in a letter to Dodgson in 1861:[16]

I have had considerable experience in dealing with minds of low logical power, and have found that studies may be made so easy and mechanical as to render thought almost superfluous.

The Euclid debate

In most English private schools, the Victorian curriculum consisted mainly of the classical languages of Latin and Greek, together with some divinity. For those schools that also taught mathematics, Euclid's geometry was the standard fare, being regarded as the ideal vehicle for teaching young men how to reason and think logically. Based on 'absolutes', the study of geometry fitted well with the classical curriculum, providing a suitable training for those expecting to go on to Oxford and Cambridge Universities and the Church. The *Elements* thus became an important constituent of examination syllabuses, being required also for entrance to the civil service and the army.

The benefits of geometry were extolled by William Whewell, Professor of Moral Philosophy at the University of Cambridge:[17]

There is no study by which the Reason can be so exactly and rigorously exercised. In learning Geometry the student is rendered familiar with the most perfect examples of strict inference…He is accustomed to a chain of deduction in which each link hangs from the preceding, yet without any insecurity in the whole: to an ascent, beginning from the solid ground, in which each step, as soon as it is made, is a foundation for the further ascent, no less solid than the first self-evident truths. Hence he learns continuity of attention, coherency of thought, and confidence in the power of human reason to arrive at the truth.

We require our present Mathematical studies not as an instrument (for the solution of today's mathematical problems) but as an exercise of the intellectual powers; that is, not for their results, but for the intellectual habits which they generate that such studies are pursued.

William Whewell and Dionysius Lardner

Moreover, Euclid's *Elements* seemed the perfect vehicle for imparting this material, as put into context by the Irish scientific popularizer Dionysius Lardner:[18]

Two thousand years have now rolled away since Euclid's Elements were first used in the school of Alexandria, and to this day they continue to be esteemed the best introduction to mathematical science. They have been adopted as the basis of geometrical instruction in every part of the globe to which the light of science has penetrated; and while in every other department of human knowledge there have been almost as many manuals as schools, in this, and in this only, one work has, by common consent, been adopted as an universal standard…

This unprecedented unanimity in the adoption of one work as the basis of instruction has not arisen from the absence of other treatises on the same subject. Some of the most eminent mathematicians have written, either original Treatises, or modifications and supposed improvements of the Elements, but still the "Elements" themselves have been invariably preferred. To what can a preference so universal be attributed, if not to that singular perspicuity of arrangement, and that rigorous exactitude of demonstration, in which this celebrated Treatise has never been surpassed?

Others, however, were opposed to the dry and over-formalistic approach of Euclid and other Greek writers. They regarded such a strictly logical approach as obscure, unsuitable for beginners, or artificial in its insistence on a minimal set of axioms.

Another objection was that the formal study of Euclid failed to encourage independent thinking, requiring too much rote learning, often with little or no understanding of geometry or of other areas of mathematics. As early as 1832 Baden Powell, Oxford's

Savilian Professor of Geometry, had complained of those students who were studying mathematics that[19]

though a certain portion had 'got up' the first four books of Euclid, not more than two or three could add Vulgar Fractions or tell the cause of day or night or the principle of a pump,

while the story was told of an Oxford examination candidate who reproduced a proof from Euclid perfectly, except that in his diagrams he drew all his triangles as circles.[20]

It was a time of change. A growing middle class was demanding a more practical approach to mathematics, extending its application to areas such as surveying, and the traditional classical education became increasingly irrelevant. In his 1869 Presidential address to the British Association for the Advancement of Science, the mathematician J. J. Sylvester was forthright in his condemnation of the old ways.[21] Regretting that

The early study of Euclid made me a hater of geometry, which I hope may plead my excuse if I have shocked the opinions of any in this room (and I know there are some who rank Euclid as second in sacredness to the Bible alone, and as one of the advanced outposts of the British constitution)

he compared the study of mathematics with that of natural science:

no one can desire more earnestly than myself to see natural and experimental science introduced into our schools as a primary and indispensable branch of education: I think that that study and mathematical culture should go on hand in hand together, and that they would greatly influence each other for their mutual good. I should rejoice to see mathematics taught with that life and animation which the presence of her young and buoyant sister could not fail to impart; short roads preferred to long ones; Euclid honourably shelved or buried 'deeper than e'er plummet sounded' out of the schoolboy's reach...

James Joseph Sylvester

Throughout the 1860s the feeling grew that examinations should no longer be based on a single book. Several texts were proposed as alternatives to the *Elements* – at first a trickle, and then a flood. As mentioned in Chapter 1, one of the most highly regarded and widely used 'rival texts', first published in 1868 and covering material from the first two Books of Euclid, was *Elements of Geometry* by J. M. Wilson, a former Cambridge Senior Wrangler who was by then teaching at Rugby School. In the preface to his book Wilson proposed the abandonment of Euclid's text in schools and colleges, on the grounds that[22]

Euclid's demonstrations are artificial, because he sacrificed simplicity and naturalness and required some unnecessary strict rules, such as the exclusion of hypothetical constructions;
the invariably syllogistic form of his reasoning is unsuitable for beginners, as it makes geometry 'unnecessarily stiff, obscure, tedious and barren';
the length of Euclid's demonstrations 'exercises the memory more than the intelligence' of the learner;
Euclid is unsuggestive as he 'places all his theorems and problems on a level, without giving prominence to the master-theorems, or clearly indicating the master-methods'.

Around this time the British Association set up an Inquiry Commission, consisting of Arthur Cayley, William Clifford, Thomas Archer Hirst, Henry Smith, George Salmon, and Sylvester himself, to investigate the whole subject of geometry teaching texts and the use of Euclid's *Elements* in particular. In an unsigned and idiosyncratic review in the *Athenaeum* of Wilson's text,[23] Augustus De Morgan reported that the Commission

has raised the question whether Euclid be, as many suppose, the best elementary treatise on geometry, or whether it be a mockery, delusion, snare, hindrance, pitfall, shoal, shallow, and snake in the grass.

De Morgan generally supported Euclid, but as a logician he realized that the deductive arguments that appeared in the *Elements* were not as flawless as its supporters made out. He continued:

We feel confidence that no system as Mr. Wilson has put forward will replace Euclid in this country. The old geometry is a very English subject, and the heretics of this orthodoxy are the extreme of heretics: even Bishop Colenso has written a Euclid. And the reason is of the same kind as that by which the classics have held their ground in education…
We only desire to avail ourselves of this feeling until the book is produced which is to supplant Euclid; we regret the manner in which it has allowed the retention of the faults of Euclid, and we trust the fight against it will rage until it ends in an amended form of Euclid.

Augustus De Morgan

On 26 May 1870, Rawdon Lovett of King Edward's School, Birmingham, proposed an Anti-Euclid Association, and by October it was able to circulate a list of twenty-eight members. This quickly re-formed itself into *The Association for the Improvement of Geometrical Teaching*, as mentioned in Chapter 1, and the new Association held its first meeting at University College, London, on 17 January 1871. Setting itself the task of producing new geometry syllabuses and texts, its interests eventually broadened to the teaching of other areas of mathematics, such as arithmetic and mechanics, while 1894 saw the establishment of its magazine, *The Mathematical Gazette*. In 1897 the Association was renamed *The Mathematical Association*, a name that survives to this day.

Another twist to this story occurred at the end of the 19th century. Although most people had believed that Euclid's geometry was the model of flawless rigorous argument, it gradually became clear that this was not the case. For example, in Euclid's construction of an equilateral triangle that we presented earlier (Book I, Proposition 1), we cannot deduce from Euclid's axioms that the two circles with centres A and B must necessarily intersect at some point C.

We have already mentioned that the logician Augustus De Morgan was unhappy with some of Euclid's arguments, and in an essay published in 1897 Bertrand Russell was even more scathing:[24]

It has been customary when Euclid, considered as a text-book, is attacked for his verbosity or his obscurity or his pedantry, to defend him on the ground that his logical excellence is transcendent, and affords an invaluable training to the youthful powers of reasoning. This claim, however, vanishes on a close inspection. His definitions do not always define, his axioms are not always indemonstrable, his demonstrations require many axioms of which he is quite unconscious. A valid proof retains its demonstrative force when no figure is drawn, but very many of Euclid's earlier proofs fail before this test.

This theme was taken up by the German mathematician David Hilbert, who in 1899 produced his *Grundlagen der Geometrie* (Foundations of Geometry), in which he presented a new set of axioms for Euclidean geometry. Based on the three primitive terms of *point*, *line*, and *plane*, and the three primitive relations of *betweenness* (point *A* lies between points *B* and *C*), *incidence* (for any points *A* and *B* there is a line that contains them both), and the *congruence* of line segments and angles, he produced twenty axioms that rigorously define Euclidean geometry from a modern point of view.

Euclid and his Modern Rivals

Charles Dodgson, an outspoken advocate for Euclid's *Elements*, was bitterly opposed to the Association for the Improvement of Geometrical Teaching and its aims. In 1879 he wrote a remarkable work, *Euclid and his Modern Rivals*, in which he skilfully compared the *Elements*, favourably in each case, with twelve well-known rival texts. A second edition appeared in 1885, to which he added a later rival text by Olaus Henrici.

Dodgson's 'modern rivals'

The rival texts that Dodgson discussed in his first edition of 1879 are, in order:

A.-M. Legendre, *Éléments de Géométrie*, 16th edition (1860)

W. D. Cooley, *The Elements of Geometry, Simplified and Explained* (1860)

Francis Cuthbertson, *Euclidian* [sic] *Geometry* (1874)

J. M. Wilson, *Elementary Geometry*, 2nd edn (1869)

Benjamin Peirce, *An Elementary Treatise on Plane and Solid Geometry* (1872)

W. A. Willock, *The Elementary Geometry of the Right Line and Circle* (1875)

W. Chauvenet, *A Treatise on Elementary Geometry* (1876)

Elias Loomis, *Elements of Geometry*, revised edn (1876)

J. R. Morell, *Euclid Simplified* (1875)

E. M. Reynolds, *Modern Methods in Elementary Geometry* (1868)

R. F. Wright, *The Elements of Plane Geometry*, 2nd edn (1871)

Syllabus of the Association for the Improvement of Geometrical Teaching (1878)

Added in the second edition of 1885 was:

Olaus Henrici, *Elementary Geometry: Congruent Figures* (1879)

Dodgson introduced his book, 'Dedicated to the memory of Euclid', as follows: [25]

The object of this little book is to furnish evidence, first, that it is essential, for the purpose of teaching or examining in elementary Geometry, to employ one textbook only; secondly, that there are strong *a priori* reasons for retaining, in all its main features, and specially in its sequence and numbering of Propositions and in its treatment of Parallels, the Manual of Euclid; and thirdly, that no sufficient reasons have yet been shown for abandoning it in favour of any one of the modern Manuals which have been offered as substitutes.

He referred particularly to the numbering of familiar results:

The Propositions have been known by those numbers for two thousand years; they have been referred to, probably, by hundreds of writers…and some of them, I. 5 and I. 47 for instance – 'the Asses' Bridge' and 'the Windmill' [the Pythagorean theorem] – are now historical characters, and their nicknames are 'familiar as household words.'

Attempting to reach a wider audience, Dodgson cast his book as a play in four acts:

It is presented in a dramatic form, partly because it seemed a better way of exhibiting in alternation the arguments on the two sides of the question; partly that I might feel myself at liberty to treat it in a rather lighter style than would have suited an essay, and thus to make it a little less tedious and a little more acceptable to unscientific readers.

There are four characters: Minos and Radamanthus (two of the three judges in Hades, appearing here as hassled Oxford examiners), Herr Niemand (the phantasm of a German professor who 'has read all books, and is ready to defend any thesis, true or untrue'), and the ghost of Euclid himself.

The curtain rises on a College study at midnight, where Minos is wearily working his way through an enormous pile of examination scripts:[26]

So, my friend! *That's* the way you prove I. 19, is it? Assuming I. 20? Cool, refreshingly cool! But stop a bit! Perhaps he doesn't 'declare to win' on Euclid. Let's see. Ah, just so! 'Legendre,' of course! Well, I suppose I must give him full marks for it: what's the question worth? – Wait a bit, though! Where's his paper of yesterday? I've a very decided impression he was all for 'Euclid' then: and I know the paper had I. 20 in it.

Ah, here it is! 'I think we do know the sweet Roman hand.' Here's the Proposition as large as life, and proved by I. 19. 'Now infidel, I have thee on the hip!' You shall have such a sweet thing to do in *vivâ-voce*, my very dear friend! You shall have the two Propositions together, and take them in any order you like. It's my profoundest conviction that you don't know how to prove either of them without the other…

These propositions from Euclid's *Elements*, Book I are:

Proposition 19: In any triangle the greater angle is subtended by the greater side.
Proposition 20: In any triangle two sides taken together in any manner are greater than the remaining one.

Minos eventually falls asleep over his scripts, but is soon awakened by the ghost of Euclid who invites him to compare the *Elements* with its Modern Rivals. To this end, Euclid summons up his friend Herr Niemand, who forcibly argues the case for each rival text. Minos carefully exposes its faults, in each case preferring Euclid's constructions, demonstrations, style, or treatment of the material. An extended extract, in which Herr Niemand and Minos discuss Cooley's book, illustrates the style of the work as a whole.[27]

Nie. I have now the honour to lay before you '*The Elements of Geometry, simplified and explained,*' by W. D. COOLEY, A.B., published in 1860.
Min. Please to hand me the book for a moment. I wish to read you a few passages from the Preface. It is always satisfactory – is it not? – to know that a writer, who attempts to 'simplify' Euclid, begins his task in a becoming spirit of humility, and with some reverence for a name that the world has accepted as an authority for two thousand years.
Nie. Truly.

MINOS *reads.*

'The Elements of Plane Geometry … are here presented in the reduced compass of 36 Propositions, perfectly coherent, fully demonstrated, and reaching quite as far as the 173 Propositions contained in the first six books of Euclid.' Modest, is it not?
Nie. A little high-flown, perhaps. Still, you know, if they really *are* 'fully demonstrated' –
Min. If! In page 4 of the Preface he talks of 'Euclid's circumlocutory shifts': in the same page he tells us that 'the doctrine of proportion, as propounded by Euclid, runs into prolixity though wanting in clearness': and again, in the same page, he states that most of Euclid's *ex absurdo* proofs 'though containing little,' yet 'generally puzzle the young student, who can hardly comprehend why gratuitous absurdities should be so formally and solemnly dealt with. These Propositions therefore are omitted from our Book of Elements, and the Problems also, for the science of Geometry lies wholly in the Theorems. Thus simplified and freed from obstructions, the truths of Geometry may, it is hoped, be easily learned, even by the youngest.' But perhaps the grandest sentence is at the end of the Preface. 'Then as to those Propositions (the first and last of the 6th Book), in which, according to the same authority' (he is alluding to the Manual of Euclid by Galbraith and Haughton), 'Euclid so beautifully illustrates his celebrated Definition, they appear to our eyes to exhibit only the verbal solemnity of a hollow logic, and to exemplify nothing but

the formal application of a nugatory principle.' Now let us see, mein Herr, whether Mr Cooley has done anything worthy of the writer of such 'brave 'orts' (as Shakespeare has it): and first let me ask how you define Parallel Lines.

NIEMAND reads.

'Right Lines are said to be parallel when they are equally and similarly inclined to the same right Line, or make equal angles with it towards the same side.'

Min. That is to say, if we see a Pair of Lines cut by a certain transversal, and are told that they make equal angles with it, we say 'these Lines are parallel'; and conversely, if we are told that a Pair of Lines are parallel, we say 'then there *is* a transversal, *somewhere*, which makes equal angles with them'?

Nie. Surely, surely.

Min. But we have no means of finding it? We have no right to draw a transversal at random and say '*this* is the one which makes equal angles with the Pair'?

Nie. Ahem! Ahem! Ahem!

Min. You seem to have a bad cough.

Nie. Let us go to the next subject.

Min. Not till you have answered my question. Have we any means of finding the particular transversal which makes the equal angles?

Nie. I am sorry for my client, but since you are so *exigeant*, I fear I must confess that we have *no* means of finding it.

Min. Now for your proof of Euc. I. 32.

Nie. You will allow us a preliminary Theorem?

Min. As many as you like.

Nie. Well, here is our Theorem II. '*When two parallel straight Lines AB, CD, are cut by a third straight Line EF, they make with it the alternate angles AGH, GHD, equal; and also the two internal angles at the same side BGH, GHD equal to two right angles.*

For *AGH* and *EGB* are equal because vertically opposite, and *EGB* is also equal to *GHD* (Definition); therefore –'

Min. There I must interrupt you. How do you know that *EGB* is equal to *GHD*? I grant you that, by the definition, *AB* and *CD* make equal angles with *a certain* transversal: but have you any ground for saying that *EF is* the transversal in question?

Nie. We have not. We surrender at discretion. You will permit us to march out with the honours of war?

Min. We grant it you of our royal grace. March him off the table, and bring on the next Rival.

After demolishing each rival book in turn, Minos is approached by Euclid, together with the phantasms of Archimedes, Pythagoras, Aristotle, Plato, &c., who have come to see fair play:

Euc. …Let us to business. And first, have you found any method of treating Parallels to supersede mine?

Min. No! A thousand times, no! The infinitesimal method, so gracefully employed by M. Legendre, is unsuited to beginners: the method by transversals, and the method by revolving Lines, have not yet been offered in a logical form: the 'equidistant' method is too cumbrous: and as for the 'method of direction', it is simply a rope of sand — it breaks to pieces wherever you touch it!

Euc We may take it as a settled thing, then, that you have found no sufficient cause for abandoning either my sequence of Propositions or their numbering…

The book concludes with Euclid's summing-up:

Euc. 'The cock doth craw, the day doth daw,' and all respectable ghosts ought to be going home. Let me carry with me the hope that I have convinced you of the importance, if not the necessity, of retaining my order and numbering, and my method of treating straight Lines, angles, right angles, and (most especially) Parallels. Leave me these untouched, and I shall look on with great contentment while other changes are made — while my proofs are abridged and improved — while alternative proofs are appended to mine — and while new Problems and Theorems are interpolated.

In all these matters my Manual is capable of unlimited improvement.

The ghosts then disappear and Minos wakes with a start and betakes himself to bed, 'a sadder and a wiser man'.

The book was a *tour de force*, exhibiting Dodgson's intimate knowledge and deep understanding of Euclidean geometry. But in spite of all his efforts it ultimately failed to achieve its aim as the tide flowed inexorably against those who wished to preserve Euclid's approach to the subject. In 1888 the Oxford and Cambridge Examination Boards, fearing that abandoning Euclid would reduce the examination systems to chaos, reluctantly agreed to accept proofs other than Euclid's, provided that they did not violate Euclid's ordering of the propositions. By 1903 this restriction too had disappeared, when they agreed to accept any systematic treatment. Indeed, a British Association committee on the teaching of elementary mathematics had already demonstrated in 1902 how far views had changed:[28]

In the opinion of the Committee, it is not necessary that one (and only one) text-book should be placed in the position of authority in demonstrative geometry; nor is it necessary that there should be only a single syllabus in control of all examinations. Each large examining body might propound its own syllabus, in the construction of which regard would be paid to the average requirements of the examiners...

In every case, the details of any syllabus should not be made too precise. It is preferable to leave as much freedom as possible, consistently with the range to be covered; for in that way the individuality of the teacher can have its most useful scope. It is the competent teacher, not the examining body, who can best find out what sequence is more suited educationally to the particular class that has to be taught.

Formal versions of the subject continued to be taught in many high schools until about the 1960s, when it was quietly dropped. Since then it has floated in and out of syllabuses, leaving few schoolchildren with an awareness of the Euclidean approach to geometry.

Euclid's 'parallel postulate'

A major difficulty with the *Elements* was Euclid's fifth postulate (or the 12th axiom in Simson's numbering) from Book I. As we saw earlier, the first four postulates are short and simple, but this one is more complicated:

If a straight line falling on two straight lines makes the interior angles on the same side less than two right angles, the two straight lines, if produced indefinitely, meet on that side on which are the angles less than the two right angles.

This says that in the figure, if angle *A* + angle *B* is less than 180°, then the two lines must eventually meet on the right, as illustrated.

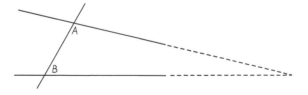

For two thousand years, generations of mathematicians tried to deduce the fifth postulate from the other four. In *Euclid and his Modern Rivals* Minos observes:[29]

Min: An absolute *proof* of it, from first principles, would be received, I can assure you, with absolute *rapture*, being an *ignis fatuus* [a delusive hope] that mathematicians have been chasing from your age down to our own.
Euc: I know it. But I cannot help you. Some mysterious flaw lies at the root of the subject.

One approach to proving the fifth postulate was to find other results that are 'equivalent' to it, in the sense that if we can prove any of these then the fifth postulate follows, and vice versa. Two such equivalent results from Euclid's *Elements*, Book I, can be written in the forms:

The sum of the angles in any triangle is 180°.
Given any line L and any point P that does not lie on this line, there is exactly one line, parallel to L, that passes through P.

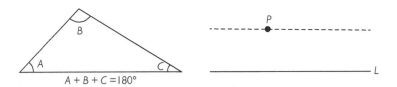

$$A + B + C = 180°$$

Because of the latter result, the fifth postulate became widely known as the *parallel postulate*. Dodgson was well aware of these equivalent versions, and in *Euclid and his Modern Rivals* he presented tables listing several more of them.[30]

In fact, it is *not* possible to deduce these results from the other postulates. This was demonstrated around 1830 by Nikolai Lobachevskii of Russia and János Bolyai of Hungary, who independently constructed strange types of geometry in which the first four postulates hold but the fifth one does not. In such 'hyperbolic geometries' the angles in any triangle add up to *less than* 180°, and given any line L and any point P that does not lie on this line there are *infinitely many* lines parallel to L that pass through P. Moreover, whenever two triangles are similar (same shape) in such a geometry, then they are necessarily congruent (same size).

Dodgson was aware of these 'non-Euclidean geometries', but rejected them as meaningless, and irrelevant to the geometrical world in which we live. He accepted that they were consistent mathematical theories, but did not recognize them as depictions of real space.[31] In 1888 he produced his volume *Curiosa Mathematica, Part I: A New Theory of Parallels*, in which he replaced Euclid's fifth postulate and its equivalent versions by a seemingly more 'obvious' one, which asserts that, given a hexagon inscribed in a circle,

the area inside the hexagon is larger than the area of any one of the six pieces that lie between the hexagon and the circle.

This version had the advantage that one can see directly that it is true – it is a 'finite' result, in contrast to Euclid's 'infinite version' in which any two non-parallel lines meet 'somewhere out there'. (This dichotomy is developed in Chapter 7.)

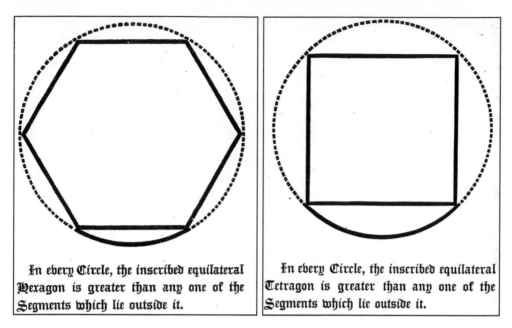

In every Circle, the inscribed equilateral Hexagon is greater than any one of the Segments which lie outside it.

In every Circle, the inscribed equilateral Tetragon is greater than any one of the Segments which lie outside it.

Two versions of Dodgson's alternative postulate

In the third and fourth editions of *A New Theory of Parallels* Dodgson replaced the hexagon by a square – in fact, any regular polygon can be used. Assuming the truth of this self-evident postulate, he deduced the fifth postulate and thereby all the other results that are equivalent to it.

Conclusion

In this final section we mention some related geometrical topics that occupied Dodgson's attention over the years.

Geometrical recreations

Among the recreational puzzles with which Dodgson entertained his adult and child friends were a number of geometrical problems. Some of these are described in Chapter 6, while others included a 'proof' that every triangle is isosceles, and a pamphlet concerning the Oxford University Parliamentary election of 1865, written in pseudo-Euclidean terms.[32] Less recreational in nature were about twenty of Dodgson's *Pillow-Problems* that involved the geometry of the plane and three-dimensional space (see Chapter 6).

Squaring the circle

Euclid's *Elements* contains a number of ruler-and-compass constructions for 'squaring' various shapes – that is, for constructing a square with the same area as the given shape. In particular, Euclid showed how to square any given triangle, quadrilateral, or pentagon.[33]

One of the classical mathematical problems of ancient Greece was to find a corresponding construction for 'squaring the circle': this problem is closely related to the properties of the number $\pi = 3.14159\ldots$, the ratio of the circle's circumference to its diameter. Over two thousand years many people believed that they had discovered such a construction, but all were wrong and in popular culture the phrase 'squaring the circle' came to mean attempting the impossible. But it was not until 1882 that a German mathematician, Ferdinand Lindemann, proved conclusively that the task really is impossible.[34]

For many years Dodgson was plagued with cranks who sent him supposed constructions for squaring the circle, or 'proofs' that π has an exact value different from the one above, such as 3, 3.125, or 3.2. In order to try to convince such enthusiasts of the errors of their ways, Dodgson began to write a book, *Simple Facts about Circle-Squaring*, that would set the record straight. In particular, he proposed to settle the argument by presenting simple arguments to convince his readers that[35]

whatever be the *exact* value of the area, it is at any rate less than 3.1417 times, and greater than 3.1413 times, the square on its radius.

Sadly, the book was never completed, and misguided enthusiasts continue to plague mathematicians around the world with their attempts to square the circle.

Right-angled triangles

One of Dodgson's last pieces of mathematics was a geometrical problem in which he sought three right-angled triangles of the same area whose sides were all whole numbers. His diary entry for 19 December 1897 reads:[36]

Sat up last night till 4 a.m., over a tempting problem, sent me from New York, "to find three equal rational-sided right-angled triangles". I found *two*, whose sides are 20, 21, 29; 12, 35, 37: but could not find *three*.

Here, 'rational-sided' means that all the sides are whole numbers or fractions, while 'equal' means that they have equal areas. Dodgson's two triangles are shown here: each has area 210.

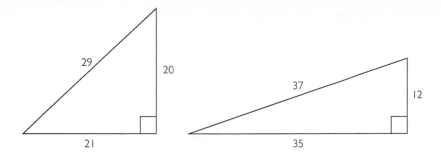

In fact, Dodgson was closer than he realized. The smallest solution of this problem in whole numbers consists of the three right-angled triangles with sides

40, 42, 58,   24, 70, 74,   and   15, 112, 113,

and common area 840; the first two of these have double the side-lengths that Dodgson found. It is now known that there are infinitely many solutions to this problem; another has sides

105, 208, 233,   120, 182, 218,   and   56, 390, 394,

and common area 10,920.

AN

ELEMENTARY TREATISE

ON

DETERMINANTS

WITH THEIR APPLICATION TO

SIMULTANEOUS LINEAR EQUATIONS

AND ALGEBRAICAL GEOMETRY.

BY

CHARLES L. DODGSON, M.A.

STUDENT AND MATHEMATICAL LECTURER OF CHRIST CHURCH, OXFORD.

London:

MACMILLAN AND CO.

1867.

Dodgson's *An Elementary Treatise on Determinants* of 1867

CHAPTER 3

Algebra

ADRIAN RICE

A famous story (flatly denied by Dodgson) relates that Queen Victoria, having enjoyed *Alice's Adventures in Wonderland*, expressed a desire to read the next book by its author, and was somewhat surprised to receive a copy of *An Elementary Treatise on Determinants*. But regardless of its truth or falsity, the attraction of this story is rooted in the huge contrast between the playful, witty, and highly readable children's novel and the cold, formal, and somewhat austere style of Dodgson's work on algebra.

Yet despite weaknesses inherent in his presentation, Dodgson's algebraic publications represent perhaps his most significant research-level contributions, and contain some interesting and original mathematics. This chapter examines his principal work on algebra – his research into determinants – discussing its possible motivation, its mathematical foundation, and its influence (or lack of it) on later mathematical developments.

Introduction

During the 19th century, algebra (like much of mathematics) was in a state of flux. For centuries its study had comprised techniques for the analysis and solution of equations, but after 1800 the subject began to move in much more abstract and theoretical directions, with the creation of innovative new algebraic entities such as groups, rings, vectors, matrices, and hypercomplex numbers.

The Mathematical World of Charles L. Dodgson (Lewis Carroll). Robin Wilson and Amirouche Moktefi.
Oxford University Press (2019). © Oxford University Press 2019.
DOI: 10.1093/oso/9780198817000.001.0001

In common with other areas of pure mathematics, algebra also gradually became more 'axiomatic', with mathematicians placing increasing importance on laying down a small set of simple but fundamental rules, or axioms, on which the entirety of the subject could then be based. At first the rules of algebra were no different from those of basic arithmetic, but it was soon discovered that consistent results could sometimes be obtained when certain previously indispensable axioms were rejected. For example, matrices do not obey the *commutative law* of multiplication: if A and B are two matrices then in general, $A \times B \neq B \times A$.

This change in algebra, from the study of equations to the much more sophisticated analysis of algebraic structures like groups and rings, led to a tremendous growth of the subject throughout the century. In Britain this was reflected in the development of the algebraic theory of invariants, principally by Arthur Cayley and James Joseph Sylvester, the creation of non-commutative algebras of quaternions and vectors by William Rowan Hamilton and Oliver Heaviside, and the introduction of algebraic symbolism into the study of logic by George Boole and Augustus De Morgan.[1]

The upshot was that the algebra of the end of the 19th century looked completely different from that of the beginning. A professor of mathematics in 1800 would have been largely familiar with the main themes and techniques of the subject of algebra, the details of which could usually be expected to fit quite comfortably in the pages of a large but single-volume textbook. By 1900 it is hard to imagine anyone, even a practising algebraist, having sufficient erudition to be conversant in all of the relevant algebraic disciplines – the subject had simply become too vast.

Dodgson's work on algebra falls about two-thirds of the way into this period of rapid development. It is almost entirely concerned with the subject of *determinants*, yet another algebraic construct that had come rapidly to the fore during the course of the 19th century.

Determinants and simultaneous linear equations

The idea of a determinant arises from two related areas of mathematics: simultaneous linear equations and analytic geometry. Suppose we have two linear equations that we want to solve simultaneously:

$$4x + 7y = 15$$
$$3x + 10y = 16.$$

The determinant is a useful device that essentially 'determines' whether these equations can be solved or not by giving a numerical value. If that number is non-zero, the system of equations has a solution; if the determinant is zero, it is insoluble.

If we construct a 2×2 matrix A from the coefficients of the above equations, we get

$$A = \begin{pmatrix} 4 & 7 \\ 3 & 10 \end{pmatrix}.$$

The *determinant* of A, written as $|A|$ or det A, is the alternating sum of the numbers all possible products of the entries in this matrix for which no two entries in a product appear in the same row or column; here, the phrase 'alternating sum' means that the terms in the sum are alternately added and subtracted, so that the alternating sum of 1, 2, 3, 4, and 5 is $1 - 2 + 3 - 4 + 5 = 3$. Thus, for the general 2×2 matrix,

$$\begin{vmatrix} a & b \\ c & d \end{vmatrix} = ad - bc.$$

So the determinant of our matrix A is

$$\det A = \begin{vmatrix} 4 & 7 \\ 3 & 10 \end{vmatrix} = (4 \times 10) - (7 \times 3) = 40 - 21 = 19,$$

which is non-zero, and so our equations have a definite solution.

To find this solution using determinants, we first write down three matrices,

$$A = \begin{pmatrix} 4 & 7 \\ 3 & 10 \end{pmatrix}, \quad A_1 = \begin{pmatrix} 15 & 7 \\ 16 & 10 \end{pmatrix}, \quad \text{and} \quad A_2 = \begin{pmatrix} 4 & 15 \\ 3 & 16 \end{pmatrix},$$

where in matrix A_1 the numbers on the right-hand side of the original simultaneous equations replace the first column of matrix A, and in matrix A_2 they replace the entries in the second column of A. We next compute the determinants of each of these matrices. We already know that det $A = 19$, and

$$\det A_1 = \begin{vmatrix} 15 & 7 \\ 16 & 10 \end{vmatrix} = (15 \times 10) - (7 \times 16) = 150 - 112 = 38,$$

$$\det A_2 = \begin{vmatrix} 4 & 15 \\ 3 & 16 \end{vmatrix} = (4 \times 16) - (15 \times 3) = 64 - 45 = 19.$$

The values of x and y are now found by the following rule:

$$x = \frac{\det A_1}{\det A} = \frac{38}{19} = 2 \quad \text{and} \quad y = \frac{\det A_2}{\det A} = \frac{19}{19} = 1.$$

In analytic geometry, the simultaneous equations

$$4x + 7y = 15$$
$$3x + 10y = 16$$

represent two straight lines in two-dimensional space, so the problem of solving for x and y is equivalent to finding the point at which these lines intersect, which is indeed the point (2, 1).

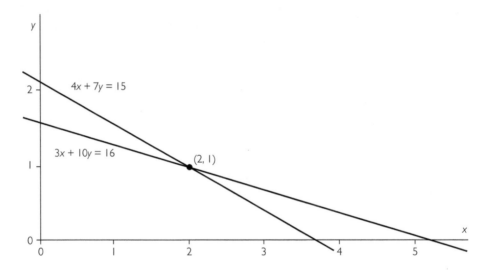

Those who have studied linear algebra may be surprised to learn that the subject of determinants actually predates that of matrices. The word *matrix*, in its current mathematical usage as a rectangular array of numbers or symbols, was coined by Sylvester in 1850, with the first major paper on the subject published by Cayley in 1858.[2] Determinants, in contrast, have a much longer history.[3] Although they can be found in the work of far-eastern mathematicians (most notably, Seki Takakazu in late 17th- and early 18th-century Japan), the modern study of determinants began with the German mathematician and philosopher Gottfried Wilhelm Leibniz in 1693, and was developed in the 18th century by Gabriel Cramer in Switzerland and Alexandre-Théophile Vandermonde in France. Indeed, the above method by which we solved the set of simultaneous equations using the division of determinants was introduced by Cramer in 1750 and is known today as *Cramer's rule.*

The next big step was made in 1772 by the French mathematician Pierre-Simon Laplace, who devised a method now known as *cofactor expansion* to work out larger determinants. For example, given the 3×3 matrix

$$B = \begin{pmatrix} 2 & 1 & 3 \\ 1 & -1 & 1 \\ 1 & 4 & -2 \end{pmatrix},$$

we can compute its determinant in the following way:

Take the entry 2, delete the rest of the first row and column, and multiply the 2 by the determinant of the remaining *minor,*

$$\begin{matrix} -1 & 1 \\ 4 & -2 \end{matrix}$$

Next move along the first row to the entry 1, delete the rest of the first row and the second column, and multiply that 1 by the determinant of the resulting minor

$$\begin{matrix} 1 & 1 \\ 1 & -2 \end{matrix}$$

Then, taking the entry 3, delete what's left of the first row and third column, and multiply the 3 by the determinant of the minor

$$\begin{matrix} 1 & -1 \\ 1 & 4 \end{matrix}$$

Finally, take the alternating sum of these three products and the result is the determinant:

$$\begin{vmatrix} 2 & 1 & 3 \\ 1 & -1 & 1 \\ 1 & 4 & -2 \end{vmatrix} = \left(2 \times \begin{vmatrix} -1 & 1 \\ 4 & -2 \end{vmatrix} \right) - \left(1 \times \begin{vmatrix} 1 & 1 \\ 1 & -2 \end{vmatrix} \right) + \left(3 \times \begin{vmatrix} 1 & -1 \\ 1 & 4 \end{vmatrix} \right)$$

$$= (2 \times -2) - (1 \times -3) + (3 \times 5) = 14.$$

Similarly, we can calculate the determinant of a 4×4 matrix by reducing it to four 3×3 determinants, and so on.

Laplace's method of cofactor expansion is effective, but it quickly becomes tedious as the dimension of the determinants increases. Computing the value of a 5 × 5 determinant involves breaking it down into five 4 × 4 determinants, which are then broken down into twenty 3 × 3 determinants, and sixty 2 × 2 determinants, requiring 120 multiplications in total and many additions and subtractions before we reach the final answer. Other techniques that were no shorter were developed by Joseph-Louis Lagrange and Carl Friedrich Gauss, via their work in number theory. In this context Gauss introduced the word *determinant* in 1801, although not in its present sense.

In 1812 the French mathematician Augustin-Louis Cauchy coined the term 'determinant' as it is currently used. Cauchy's work over the next few years extended and generalized the subject considerably, marking the beginning of a proper theory. This theory was then built on and expanded in the 1830s and 1840s by the Prussian mathematician Carl Gustav

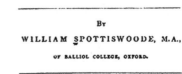

ELEMENTARY THEOREMS

RELATING TO

DETERMINANTS.

By
WILLIAM SPOTTISWOODE, M.A.,
OF BALLIOL COLLEGE, OXFORD.

LONDON : LONGMAN, BROWN, GREEN, AND LONGMAN, PATERNOSTER ROW,
AND GEORGE BELL, 186, FLEET STREET.
OXFORD : J. H. PARKER.
CAMBRIDGE : JOHN DEIGHTON.

1851.

William Spottiswoode and his book on determinants

Jacobi, whose work had a significant influence on subsequent research by British mathematicians, principally Sylvester and Cayley.

It was also reflected in the content of the first textbook on the subject, *Elementary Theorems Relating to Determinants*, published in 1851 while Dodgson was an undergraduate at Christ Church. Despite its title, this 63-page survey was a far from rudimentary treatment, containing a significant amount of material on recently published research. Its author was William Spottiswoode, an Oxford-trained mathematician and physicist, later to become President of the Royal Society. As one of Britain's foremost experts on determinants, he became a valuable source of information and advice for Dodgson during his researches into the subject.

Dodgson's motivation

Dodgson would not have come across determinants as part of the Oxford undergraduate mathematics curriculum. Spottiswoode's textbook was not aimed at students studying for university degrees, but rather for mathematicians who needed an introduction to the subject in order to read research-level publications and perhaps to carry out their own subsequent investigations. As a college lecturer in the 1850s and 1860s, Dodgson would not have been required to teach determinants or to undertake research on the subject.[4] So before examining his work on determinants in detail, we need to consider why he undertook such a study. Such a question may at first seem trivial: after all, what is unusual or remarkable about a mathematician such as Dodgson carrying out research into what was then a current area of algebraic interest?

But the question *is* of importance because, unlike many of his mathematical contemporaries such as Cayley, Sylvester, or De Morgan, it was not Dodgson's habit to carry out and publish research into mainstream mathematical topics on a regular basis. In fact, his work on determinants was the only example of any mathematics he produced that was published in a refereed scientific journal. He belonged to no academic or scientific societies, and indeed, when the London Mathematical Society was set up in 1865 as the first learned body of its kind to be established in Britain, he took no interest in its activities. Even when Sylvester established the Oxford Mathematical Society closer to home in 1888, Dodgson did not become a member – even though three of his Christ Church colleagues joined. Although employed as a college lecturer in Oxford, his duties extended no further than the teaching and pastoral care of his undergraduates. The formal obligation for university and college lecturers to engage in active research did not emerge until the 20th century, but no such requirement existed in Dodgson's day. In short, we may assume that Dodgson was inspired to conduct research into determinants because they interested him and not for any professional or financial reasons. But how did that interest arise?

Throughout his academic career, one of Dodgson's abiding interests was logical precision in reasoning, mathematical or otherwise. As Amirouche Moktefi recently pointed out:[5]

[Dodgson's] ideal of rigour that he championed in public debates unsurprisingly is also essential in his mathematical investigations. His numerous writings in this discipline attest a great concern with the precision of notations and the rigour of arguments.

As we will see in Chapter 4, this fascination with logic and rigour was rooted in his love of geometry, and in his unalterable belief in the supremacy of Euclid's *Elements* as a textbook for teaching the subject. It is thus not surprising that in his *Elementary Treatise on Determinants*, he presented the subject[6]

as a continuous chain of argument, separated from all accessories of explanation or illustration, a form which I venture to think better suited for a treatise on exact science than the semi-colloquial semi-logical form often adopted by Mathematical writers.

But while Dodgson's strong desire to place determinants on a firm logical foundation may have provided him with the impetus to devote a book-length work to the matter, it still does not explain his reasons for choosing the subject of determinants in the first place. So, we might ask, since there were many other algebraic (and mathematical) areas then in need of a more rigorous axiomatic foundation, why did determinants particularly warrant Dodgson's attention?

While his teaching may have provided an impetus, he is unlikely to have been motivated by the need to teach determinants as a subject in its own right. Whereas related topics such as solving equations had been on the Oxford mathematics curriculum for many years, the first questions on determinants did not appear on Oxford examination papers until the 1870s,[7] around the same time as they were being introduced into mathematics syllabuses elsewhere in the country.[8] Moreover, 'the concept of a matrix as something beyond a determinant was not something that ever appeared in the Oxford exams in the nineteenth century'.[9] So it seems that in thus writing an introductory book on the subject in the 1860s, prior to its introduction into the Oxford curriculum, Dodgson was either anticipating a curricular development that he viewed as highly likely, or was aware that the inclusion of determinants had already been decided, or was gifted with remarkable prescience. But whatever the case, as an increasingly important area of algebra at the time, the subject did not appear from nowhere, and with very few textbooks on the topic available (and only one in English), there was clearly a gap in the market for an 'elementary treatise'.

One further source of possible inspiration for his interest in determinants should not be discounted: his interest in what was then called 'algebraical geometry' (known today as 'analytic geometry'). Like the subject of solving equations, the topic of using algebra to solve geometric problems, or 'Cartesian geometry' as it is often called, had been a staple of the Oxford mathematics curriculum for decades. Indeed, in 1860 Dodgson had published a pamphlet entitled *A Syllabus of Plane Algebraical Geometry* to assist his students in their study of precisely this subject.

As we have seen, determinants are useful in the study of the possible intersections of lines in two-dimensional space, and this use extends to the study of planes in three-dimensional space, as well as their counterparts in higher dimensions. With the installation in 1861 of Henry Smith, Oxford's new and progressive Savilian Professor of Geometry, more questions began to appear on examination papers which, 'while geometric in setting, required an almost entirely algebraic approach'.[10] The use of determinants would thus have become increasingly important to students in their geometrical studies, and it is surely no coincidence that the final two chapters of Dodgson's 1867 book are devoted to the application of determinants to 'algebraical geometry'. But in the absence of any definitive statement on the matter from Dodgson, we will probably never know what provided the impetus for his initial interest in the subject.

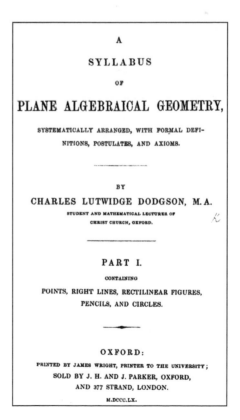

A

SYLLABUS

OF

PLANE ALGEBRAICAL GEOMETRY,

SYSTEMATICALLY ARRANGED, WITH FORMAL DEFI-
NITIONS, POSTULATES, AND AXIOMS.

BY

CHARLES LUTWIDGE DODGSON, M.A.
STUDENT AND MATHEMATICAL LECTURER OF
CHRIST CHURCH, OXFORD.

PART I.

CONTAINING

POINTS, RIGHT LINES, RECTILINEAR FIGURES,
PENCILS, AND CIRCLES.

OXFORD:
PRINTED BY JAMES WRIGHT, PRINTER TO THE UNIVERSITY;
SOLD BY J. H. AND J. PARKER, OXFORD,
AND 377 STRAND, LONDON.
M.DCCC.LX.

Dodgson's *A Syllabus of Plane Algebraical Geometry*

Dodgson and determinants

Dodgson's work on determinants was largely confined to a single two-year period of his life, which followed the completion of his first *Alice* book. Through information gleaned from his diaries and surviving excerpts from his correspondence, we know that he was actively working on the subject between the autumn of 1865 and the end of 1867, and with the aid of these sources we are able, at least partially, to trace the progress of his work. Although we cannot tell when his interest in determinants was first aroused, it is clear from his diary that towards the end of 1865 concentrated work on the subject was under way. On 28 October of that year, he recorded:[11]

I have been at work for some days on an elementary pamphlet on Determinants which I think of printing.

At this point there was no indication of any original research activity on Dodgson's part: indeed, it is likely that he first envisaged his pamphlet as being little more than an expository or pedagogical publication. But four months later, over a short but intense two-day period, Dodgson made two successive discoveries that turned out to be novel contributions to the subject. On 27 February 1866 he wrote that he had[12]

Discovered a process for evaluating arithmetical Determinants, by a sort of condensation, and proved it up to 4^2 terms.

His diary entry for the following day then observed that he had[13]

Completed a rule, built on the process discovered yesterday, for solving simultaneous simple equations. It is far the shortest method I have yet seen.

These two original discoveries seem to have changed the direction of his work, motivating him to seek advice on whether it might be worthwhile to publish them in a learned journal – something that Dodgson, up to that point, had never done.

The first person to whom he turned was his Oxford colleague and mentor Bartholomew Price, the Sedleian Professor of Natural Philosophy. The two men had been acquainted since Dodgson's undergraduate days and, according to Robin Wilson, they went on to form a lifelong mathematical friendship,[14]

with Dodgson frequently asking Price for advice on various matters, especially in his early years as Mathematical lecturer at Christ Church, and the two of them regularly collaborating on the setting of University examinations and the award of mathematical studentships.

Bartholomew Price, Sedleian Professor of Natural Philosophy

Since he was not an expert on determinants, Price sent an enquiry to Spottiswoode who, as the author of the only English book on the subject to date, was a recognized authority. By 25 March Dodgson had received a letter from Spottiswoode[15]

(to whom Price had sent my question as to the shortest way of computing Determinants arithmetically), saying that he knows of no short way, and that he would be very glad to hear from me.

Four days later, he sent Spottiswoode a written account of his new 'condensation' method for computing determinants and of its application to the solution of simultaneous linear equations.[16] Spottiswoode's response was swift and enthusiastic. A letter dated 2 April 1866 finds him writing warmly to Dodgson:[17]

Your method of computation, condensation is very successful. The Theorem upon which it is founded is, as you are doubtless aware, known; but the application of it is, as far as I am aware, quite original. I congratulate you upon it.

With Spottiswoode's endorsement, it was decided that Dodgson should write up his results into a formal paper to be submitted to no less a scientific body than the Royal Society, of which both Price and Spottiswoode were Fellows. Although Dodgson was not himself a member, Price agreed to communicate the paper on his behalf,[18] and on 17 May 1866 the paper, entitled 'Condensation of Determinants, being a new and brief method for computing their arithmetical values' was officially read to the Society. It was published soon after in the Royal Society's *Proceedings*.[19]

[*From the* PROCEEDINGS OF THE ROYAL SOCIETY, No. 84, 1866.]

CONDENSATION OF DETERMINANTS,

BEING A

NEW AND BRIEF METHOD

FOR

COMPUTING THEIR ARITHMETICAL VALUES.

BY THE

REV. C. L. DODGSON, M.A.,

STUDENT OF CHRIST CHURCH, OXFORD.

IV. "Condensation of Determinants, being a new and brief Method for computing their arithmetical values." By the Rev. C. L. DODGSON, M.A., Student of Christ Church, Oxford. Communicated by the Rev. BARTHOLOMEW PRICE, M.A., F.R.S. Received May 15, 1866.

If it be proposed to solve a set of n simultaneous linear equations, not being all homogeneous, involving n unknowns, or to test their compatibility when all are homogeneous, by the method of determinants, in these, as well as in other cases of common occurrence, it is necessary to compute the arithmetical values of one or more determinants—such, for example, as

$$\begin{vmatrix} 1, & 3, & -2 \\ 2, & 1, & 4 \\ 3, & 5, & -1 \end{vmatrix}.$$

Now the only method, so far as I am aware, that has been hitherto employed for such a purpose, is that of multiplying each term of the first row or column by the determinant of its complemental minor, and affecting the products with the signs + and − alternately, the determinants required in the process being, in their turn, broken up in the same manner until determinants are finally arrived at sufficiently small for mental computation. This process, in the above instance, would run thus:—

$$\begin{vmatrix} 1, & 3, & -2 \\ 2, & 1, & 4 \\ 3, & 5, & -1 \end{vmatrix} = 1 \times \begin{vmatrix} 1, & 4 \\ 5, & -1 \end{vmatrix} -2 \times \begin{vmatrix} 3, & -2 \\ 5, & -1 \end{vmatrix} +3 \times \begin{vmatrix} 3, & -2 \\ 1, & 4 \end{vmatrix}$$

$$= -21 - 14 + 42 = 7.$$

But such a process, when the block consists of 16, 25, or more terms, is so tedious that the old method of elimination is much to be preferred for solving simultaneous equations; so that the new method, excepting for equations containing 2 or 3 unknowns, is practically useless.

The new method of computation, which I now proceed to explain, and for which "Condensation" appears to be an appropriate name, will be found, I believe, to be far shorter and simpler than any hitherto employed.

Dodgson's Royal Society paper, 'Condensation of Determinants'

A meeting of the Fellows of the Royal Society at Somerset House in the 1840s

[Engraving by Henry Melville, *London Interiors: A Grand National Exhibition*, London, 1844]

Presumably spurred on by the reception of his paper, Dodgson resumed work on his pamphlet, which over the course of the next few months grew gradually into a small book. By February 1867 Dodgson was again requesting advice from Spottiswoode, asking for comments on proof sheets for the forthcoming volume. But in addition to various recommendations, such as giving advice to include some worked examples for the benefit of student readers, Spottiswoode was critical of Dodgson's new and unorthodox notation and terminology:[20]

If the expressions to wh[ich] they refer were new, I am not sure but that I sh[oul]d prefer yours; but it seemed to me undesirable to introduce new terms,– or to modify terms,– unless some considerable advantage was to be obtained.

Spottiswoode closed with a humorous jibe at Sylvester, well known at the time for introducing a whole host of new and sometimes bizarre mathematical nomenclature:[21]

If you ever read Sylvester's papers you will realise my horror of new terminology.

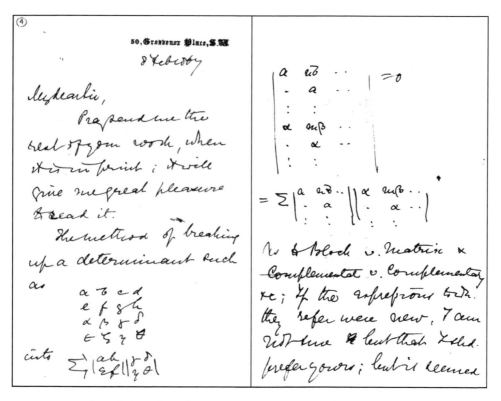

Part of a letter from Spottiswoode to Dodgson, dated 8 February 1867, concerning Dodgson's work on determinants

[Morris L. Parrish Collection, Princeton University]

Dodgson chose to ignore Spottiswoode's advice about the use of new notation, as we shall see, but he did adopt the suggestion to include worked examples in the book – although they were added merely as footnotes and appendices, so as to avoid disturbing the logical flow of the main text. Work had now progressed sufficiently far for Dodgson to write the following letter to his publisher, Alexander Macmillan, a mere three days after the letter from Spottiswoode:[22]

February II, 1867

Dear Mr. Macmillan,

I have got a little book, near completion, which I want you to publish for me – *not* one, I am afraid, that can be brought out as 'by the author of *Alice's Adventures*.' Its title is 'The Elements of Determinants, and of Their Application to Simultaneous Equations and to Algebraical Geometry – for the use of beginners.' I am having it printed at the University Press, and fully expect that it will be ready to bring out in another month or so – in good time for next term. It is octavo and, as far as I can guess, will be about 100 pages – but the selling price we can settle hereafter: no price would repay me, I fancy, as I have made so many alterations. The reason I mention it now is, that you may be able to begin advertising it whenever you think fit.

I feel quite undecided as to how many copies to print, at present. That the book is much wanted there can be no doubt: that *my* book will meet the want, may well be doubted – but I am encouraged by the favourable opinion of Mr. Spottiswoode, who is I believe *the* authority on the new subject of Determinants and who has seen most of the book. If you are willing to undertake the publication, I will write further. Perhaps you may be able to suggest what sort of number it would be wise to print – also how you would propose to bind it. Should you recommend green cloth and gold lettering, as I see you do for smaller mathematical books?

Yours truly,
C. L. Dodgson

But the book was not yet finished, and it is clear from a diary entry in mid-March that Dodgson was finding the task difficult:[23]

Went through a grand piece of 'lost labour' in my book on Determinants. At II at night I came to a difficulty which seemed to require the re-wording of most of the propositions in the chapter on Equations. I went right through the chapter, and did not get it done till about quarter past 4 in the morning, and the next day I found quite another way out of the difficulty, which made all the alterations unnecessary! This little book (it will be about 100 pages I should think) has given me more trouble than anything I have ever written: it is such entirely new ground to explore.

At the close of the academic year Dodgson took a break from his mathematics (as we mentioned in Chapter 1) when he spent the long summer vacation of 1867 travelling to Russia and Eastern Europe with his friend Henry Liddon. Returning to Oxford in October,[24] he managed to complete the book by 15 November;[25] it was finally published on 10 December 1867 under the title *An Elementary Treatise on Determinants, with their Application to Simultaneous Linear Equations and Algebraical Geometry*.[26] At 152 pages it is his only book-length publication on algebra.

Dodgson's method of condensation

The most original feature of Dodgson's work on determinants was the subject of his 1866 paper – his 'condensation' method for evaluating determinants, and its application to the solution of simultaneous linear equations. This material was also included in Appendices II–IV of his *Elementary Treatise on Determinants*.

The principal advantage of Dodgson's method becomes obvious when compared with other techniques. Recall, for example, that in Laplace's method of cofactor expansion, a 5×5 determinant needs to be decomposed first into 4×4 determinants, then a greater number of 3×3 determinants, and so on until the answer is reached. This is both laborious and complicated. However, the genius of Dodgson's method of condensation is that, regardless of the size of the matrix in question, it never requires the computation of anything more sophisticated than 2×2 determinants.

Its procedure can best be illustrated with a couple of examples. We have already seen that the determinant of the following 3×3 matrix is 14:

$$B = \begin{pmatrix} 2 & 1 & 3 \\ 1 & -1 & 1 \\ 1 & 4 & -2 \end{pmatrix}.$$

But Dodgson's method gives us an easier way to find this answer. We begin by simply calculating the determinant of each of the four 2×2 connected minors within the matrix – that is, we compute

$$\begin{vmatrix} \begin{vmatrix} 2 & 1 \\ 1 & -1 \end{vmatrix} & \begin{vmatrix} 1 & 3 \\ -1 & 1 \end{vmatrix} \\ \begin{vmatrix} 1 & -1 \\ 1 & 4 \end{vmatrix} & \begin{vmatrix} -1 & 1 \\ 4 & -2 \end{vmatrix} \end{vmatrix} = \begin{vmatrix} -3 & 4 \\ 5 & -2 \end{vmatrix},$$

which gives the apparently incorrect answer of –14. But the method has one final step, which is to divide this result by the entry (–1) in the middle of the original matrix. This leads to the true value of 14.

In the more sophisticated case of a 4 × 4 matrix, such as

$$C = \begin{pmatrix} 2 & 0 & 3 & 1 \\ 1 & 4 & 2 & 3 \\ 0 & 3 & 1 & 2 \\ 3 & 2 & 0 & 1 \end{pmatrix},$$

the procedure is similar. Once again, this matrix is 'condensed' down to a 3 × 3 matrix,

$$\begin{vmatrix} \begin{vmatrix} 2 & 0 \\ 1 & 4 \end{vmatrix} & \begin{vmatrix} 0 & 3 \\ 4 & 2 \end{vmatrix} & \begin{vmatrix} 3 & 1 \\ 2 & 3 \end{vmatrix} \\ \begin{vmatrix} 1 & 4 \\ 0 & 3 \end{vmatrix} & \begin{vmatrix} 4 & 2 \\ 3 & 1 \end{vmatrix} & \begin{vmatrix} 2 & 3 \\ 1 & 2 \end{vmatrix} \\ \begin{vmatrix} 0 & 3 \\ 3 & 2 \end{vmatrix} & \begin{vmatrix} 3 & 1 \\ 2 & 0 \end{vmatrix} & \begin{vmatrix} 1 & 2 \\ 0 & 1 \end{vmatrix} \end{vmatrix} = \begin{vmatrix} 8 & -12 & 7 \\ 3 & -2 & 1 \\ -9 & -2 & 1 \end{vmatrix},$$

which is then 'condensed' to

$$\begin{vmatrix} \begin{vmatrix} 8 & -12 \\ 3 & -2 \end{vmatrix} & \begin{vmatrix} -12 & 7 \\ -2 & 1 \end{vmatrix} \\ \begin{vmatrix} 3 & -2 \\ -9 & -2 \end{vmatrix} & \begin{vmatrix} -2 & 1 \\ -2 & 1 \end{vmatrix} \end{vmatrix} = \begin{vmatrix} 20 & 2 \\ -24 & 0 \end{vmatrix}.$$

At this stage, we divide each term in this 2 × 2 determinant by its corresponding entry in the middle of the original 4 × 4 matrix (marked in bold type in the matrix C), to give

$$\begin{vmatrix} 5 & 1 \\ -8 & 0 \end{vmatrix} = 8.$$

Finally, this last number is divided by the bold entry (–2) in the middle of the preceding 3 × 3 matrix, to give $\det C = -4$.

While this is all very ingenious, this method gives us little insight into why it works. However, as Spottiswoode had pointed out, it is based on a well-known theorem in determinants, which Dodgson presented (in slightly different forms) in both his paper and his subsequent book. This theorem was first stated and proved by Jacobi in

$1833,$[27] although special cases of it had been derived independently by earlier mathematicians.

In order to understand it, we need two definitions, the first of which is a matrix that Dodgson termed the 'adjugate'. Given an $n \times n$ matrix

$$M = \begin{pmatrix} a_{11} & a_{12} & \cdots & a_{1n} \\ a_{21} & a_{22} & \cdots & a_{2n} \\ \vdots & \vdots & \ddots & \vdots \\ a_{n1} & a_{n2} & \cdots & a_{nn} \end{pmatrix},$$

its *adjugate* M' is defined to be

$$M' = \begin{pmatrix} a'_{11} & a'_{12} & \cdots & a'_{1n} \\ a'_{21} & a'_{22} & \cdots & a'_{2n} \\ \vdots & \vdots & \ddots & \vdots \\ a'_{n1} & a'_{n2} & \cdots & a'_{nn} \end{pmatrix},$$

where each entry $a'_{ij} = (-1)^{i+j} \det\left[M_{ij}\right]$ and $\left[M_{ij}\right]$ is the minor obtained by deleting row i and column j from M. As an example, given the matrix

$$B = \begin{pmatrix} 2 & 1 & 3 \\ 1 & -1 & 1 \\ 1 & 4 & -2 \end{pmatrix},$$

its adjugate is

$$B' = \begin{pmatrix} \begin{vmatrix} -1 & 1 \\ 4 & -2 \end{vmatrix} & -\begin{vmatrix} 1 & 1 \\ 1 & -2 \end{vmatrix} & \begin{vmatrix} 1 & -1 \\ 1 & 4 \end{vmatrix} \\ -\begin{vmatrix} 1 & 3 \\ 4 & -2 \end{vmatrix} & \begin{vmatrix} 2 & 3 \\ 1 & -2 \end{vmatrix} & -\begin{vmatrix} 2 & 1 \\ 1 & 4 \end{vmatrix} \\ \begin{vmatrix} 1 & 3 \\ -1 & 1 \end{vmatrix} & -\begin{vmatrix} 2 & 3 \\ 1 & 1 \end{vmatrix} & \begin{vmatrix} 2 & 1 \\ 1 & -1 \end{vmatrix} \end{pmatrix} = \begin{pmatrix} -2 & 3 & 5 \\ 14 & -7 & -7 \\ 4 & 1 & -3 \end{pmatrix}.$$

The second definition is more straightforward. Dodgson defined the *interior*, int M, of an $n \times n$ matrix M to be the $(n-2) \times (n-2)$ matrix that results from deleting the first and last rows and columns of M; for the matrices B and C above,

$$\text{int } B = (-1) \quad \text{and} \quad \text{int } C = \begin{pmatrix} 4 & 2 \\ 3 & 1 \end{pmatrix}.$$

To shed light on the workings of Dodgson's algorithm, we use a particular case of Jacobi's theorem,[28] which says that, if M is an $n \times n$ matrix, and if

$$\begin{pmatrix} a'_{11} & a'_{1n} \\ a'_{n1} & a'_{nn} \end{pmatrix}$$

is the 2×2 minor consisting of the corner entries of its adjugate M', then

$$\det \begin{pmatrix} a'_{11} & a'_{1n} \\ a'_{n1} & a'_{nn} \end{pmatrix} = \det \left(\text{int } M \right) \times \det M. \tag{*}$$

Dodgson had clearly noticed that if the determinant of M is unknown, then Jacobi's theorem gives a useful algorithm for finding it. Suppose, for example, that we do not know that $\det B = 14$. We know, from having computed its adjugate B' above, that

$$\begin{pmatrix} a'_{11} & a'_{13} \\ a'_{31} & a'_{33} \end{pmatrix} = \begin{pmatrix} -2 & 5 \\ 4 & -3 \end{pmatrix},$$

and the determinant of this 2×2 matrix is -14. Now by Jacobi's theorem (*), this value must be equal to $\det (\text{int } B) \times \det B = -\det B$, because $\text{int } B = -1$. So $-\det B = -14$ and $\det B = 14$. This gives a simple illustration, in the 3×3 case, of why Dodgson's condensation method works.[29]

But despite the appeal of this method, its principal weakness is that it relies on dividing by numbers in the interior of a matrix. If this interior contains zeros, the method breaks down. For example, applying Dodgson's method to the matrix

$$D = \begin{pmatrix} 2 & -1 & 3 & 1 \\ 1 & 2 & 1 & 1 \\ 0 & 2 & 1 & 2 \\ 3 & 2 & -1 & 1 \end{pmatrix}$$

gives

$$\begin{vmatrix} 5 & -7 & 2 \\ 2 & 0 & 1 \\ -6 & -4 & 3 \end{vmatrix} \rightarrow \begin{vmatrix} 14 & -7 \\ -8 & 4 \end{vmatrix} \rightarrow \begin{vmatrix} \dfrac{14}{2} & \dfrac{-7}{1} \\ \dfrac{-8}{2} & \dfrac{4}{1} \end{vmatrix} \rightarrow \begin{vmatrix} 7 & -7 \\ -4 & 4 \end{vmatrix} \rightarrow \dfrac{0}{0},$$

which is undefined.

In this case, however, there is a remedy. As Dodgson noticed, we may be able to prevent zeros from appearing in the interior by simply interchanging rows or columns.

Since the determinant is unaffected when an *even* number of rows (or columns) are switched, such a tactic can sometimes be used to eliminate the difficulty. For the matrix D, interchanging row 1 with row 3 and row 2 with row 4 resolves the difficulty and det D is easily found to be 30. But this remedy does not work in every case, exposing the main practical weakness of the condensation method. For this reason Dodgson noted, almost apologetically, that he regarded it[30]

merely as a fanciful addition to the processes already in use, which may in some cases lessen the labour of computation.

As an instance of the foregoing rules, let us take the block

$$\begin{vmatrix} -2 & -1 & -1 & -4 \\ -1 & -2 & -1 & -6 \\ -1 & -1 & 2 & 4 \\ 2 & 1 & -3 & -8 \end{vmatrix}.$$

By rule (2) this is condensed into $\begin{vmatrix} 3 & -1 & 2 \\ -1 & -5 & 8 \\ 1 & 1 & -4 \end{vmatrix}$; this, again, by rule (3), is condensed into $\begin{vmatrix} 8 & -2 \\ -4 & 6 \end{vmatrix}$; and this, by rule (4), into -8, which is the required value.

The simplest method of working this rule appears to be to arrange the series of blocks one under another, as here exhibited; it will then be found very easy to pick out the divisors required in rules (3) and (4).

$$\begin{vmatrix} -2 & -1 & -1 & -4 \\ -1 & -2 & -1 & -6 \\ -1 & -1 & 2 & 4 \\ 2 & 1 & -3 & -8 \end{vmatrix}$$

$$\begin{vmatrix} 3 & -1 & 2 \\ -1 & -5 & 8 \\ 1 & 1 & -4 \end{vmatrix}$$

$$\begin{vmatrix} 8 & -2 \\ -4 & 6 \end{vmatrix}$$

$$-8.$$

Dodgson illustrates his method of condensation

Dodgson was nevertheless quick to offer an immediate application of the method, to the solution of systems of simultaneous linear equations. Given such a system,

$$\begin{aligned} 2x_2 + x_3 + 2x_4 &= -3 \\ 3x_1 + 2x_2 - x_3 + x_4 &= 7 \\ 2x_1 - x_2 + 3x_3 + x_4 &= -2 \\ x_1 + 2x_2 + x_3 + x_4 &= 1, \end{aligned}$$

$$\begin{array}{cccccc}- & + & - & + & - & +\end{array}$$

$$
\begin{aligned}
x &+2y + z - u +2v + 2 = 0\\
x &- y -2z \;\;\;\; u - v - 4 = 0\\
2x &+ y - z -2u - v - 6 = 0\\
x &-2y - z - u +2v + 4 = 0\\
2x &- y +2z + u -3v - 8 = 0
\end{aligned}
$$

$$
\begin{vmatrix}
1 & 2 & 1 & -1 & 2 & 2\\
1 & -1 & -2 & -1 & -1 & -4\\
2 & 1 & -1 & -2 & -1 & -6\\
1 & -2 & -1 & -1 & 2 & 4\\
2 & -1 & 2 & 1 & -3 & -8
\end{vmatrix}
\quad
\begin{vmatrix}
2 & 4\\
-1 & -2\\
-1 & -2\\
2 & 6
\end{vmatrix}
\quad
\begin{vmatrix}
2 & 6\\
-1 & -3\\
-1 & -1\\
3 & 0
\end{vmatrix}
\quad
\begin{vmatrix}
2 & 5\\
-1 & -1\\
3 & 3
\end{vmatrix}
\quad
\begin{vmatrix}
2 & 4
\end{vmatrix}
$$

$\therefore -2v=4$
$\therefore v=-2$
$\therefore 3u=3 \;\ldots\ldots\; \therefore u= 1$

$$
\begin{vmatrix}
-3 & -3 & -3 & 3 & -6\\
3 & 3 & 3 & -1 & 2\\
-5 & -3 & -1 & -5 & 8\\
3 & -5 & 1 & 1 & -4
\end{vmatrix}
\quad
\begin{vmatrix}
-1 & 0\\
-5 & -2
\end{vmatrix}
\quad
\begin{vmatrix}
6 & 6
\end{vmatrix}
$$

$\therefore -6z=6 \ldots\ldots\ldots\ldots\ldots\ldots \therefore z=-1$

$$
\begin{vmatrix}
6 & 0\\
8 & -2
\end{vmatrix}
$$

$$
\begin{vmatrix}
0 & 0 & 6 & 0\\
6 & -6 & 8 & -2\\
-17 & 8 & -4 & 6
\end{vmatrix}
\quad
\begin{vmatrix}
12 & 12
\end{vmatrix}
$$

$\therefore 12y=12 \ldots\ldots\ldots\ldots\ldots\ldots \therefore y= 1$

$$
\begin{vmatrix}
0 & 12 & 12\\
18 & 40 & -8
\end{vmatrix}
$$

$$
\begin{vmatrix}
36 & -72
\end{vmatrix}
$$

$\because -36x=-72 \ldots\ldots\ldots\ldots\ldots\ldots \therefore x= 2$

$$\begin{array}{cccc}- & + & - & +\end{array}$$

$$
\begin{aligned}
5x &+2y -3z + 3 = 0\\
3x &- y -2z + 7 = 0\\
2x &+3y + z -12 = 0
\end{aligned}
$$

$$
\begin{vmatrix}
5 & 2 & -3 & 3\\
3 & -1 & -2 & 7\\
2 & 3 & 1 & -12
\end{vmatrix}
\quad
\begin{vmatrix}
-3 & 8\\
-2 & 10
\end{vmatrix}
\quad
\begin{vmatrix}
-3 & 12
\end{vmatrix}
$$

$\therefore 3z=12 \ldots\ldots\ldots \therefore z= 4$

$$
\begin{vmatrix}
-11 & -7 & -15\\
11 & 5 & 17
\end{vmatrix}
\quad
\begin{vmatrix}
-7 & -14
\end{vmatrix}
$$

$\therefore -7y=-14 \ldots\ldots\ldots\ldots \therefore y= 2$

$$
\begin{vmatrix}
-22 & 22
\end{vmatrix}
$$

$\therefore 22x=22 \ldots\ldots\ldots\ldots \therefore x= 1$

Dodgson's use of condensation to solve a more complicated system of linear equations, from his *Elementary Treatise on Determinants*

Dodgson would construct the associated 'augmented' 4 × 5 matrix, which also includes the right-hand side of the system of equations:

$$\begin{pmatrix} 0 & 2 & 1 & 2 & -3 \\ 3 & 2 & -1 & 1 & 7 \\ 2 & -1 & 3 & 1 & -2 \\ 1 & 2 & 1 & 1 & 1 \end{pmatrix}.$$

Within this matrix are two 4 × 4 matrices,

$$E = \begin{pmatrix} 0 & 2 & 1 & 2 \\ 3 & 2 & -1 & 1 \\ 2 & -1 & 3 & 1 \\ 1 & 2 & 1 & 1 \end{pmatrix} \quad \text{and} \quad E_1 = \begin{pmatrix} 2 & 1 & 2 & -3 \\ 2 & -1 & 1 & 7 \\ -1 & 3 & 1 & -2 \\ 2 & 1 & 1 & 1 \end{pmatrix},$$

the second of which is essentially matrix E with the first column replaced by the numbers on the right-hand sides of the equations and the positions of the columns changed. (Note that since an odd number of columns have been interchanged, the sign of the determinant of E_1 is the negative of what it would have been had the positions of the columns stayed the same.) By Cramer's rule, which we introduced earlier, we find the value of the unknown x_1 to be

$$x_1 = -\frac{\det E_1}{\det E}.$$

These determinants are easily calculated by the condensation method to give $\det E = 30$ and $\det E_1 = -60$, so that $x_1 = 2$. A similar method can be used to find the other unknowns. Alternatively, substituting $x_1 = 2$ back into the original equations gives a simpler system:

$$2x_2 + x_3 + 2x_4 = -3$$
$$2x_2 - x_3 + x_4 = 1$$
$$-x_2 + 3x_3 + x_4 = -6$$
$$2x_2 + x_3 + x_4 = -1.$$

Since we are now looking for three unknowns, we need only three equations. If we ignore the last equation, we obtain the following 3 × 4 matrix:

$$\begin{pmatrix} 2 & 1 & 2 & -3 \\ 2 & -1 & 1 & 1 \\ -1 & 3 & 1 & -6 \end{pmatrix}.$$

This gives rise to two 3×3 matrices:

$$F = \begin{pmatrix} 2 & 1 & 2 \\ 2 & -1 & 1 \\ -1 & 3 & 1 \end{pmatrix} \quad \text{and} \quad F_1 = \begin{pmatrix} 1 & 2 & -3 \\ -1 & 1 & 1 \\ 3 & 1 & -6 \end{pmatrix},$$

and a combination of Cramer's rule and Dodgson's condensation gives

$$x_2 = \frac{\det F_1}{\det F} = \frac{-1}{-1} = 1.$$

Continuing in this manner produces the 2×2 matrices

$$G = \begin{pmatrix} 1 & 2 \\ -1 & 1 \end{pmatrix} \quad \text{and} \quad G_1 = \begin{pmatrix} 2 & -5 \\ 1 & -1 \end{pmatrix},$$

so that

$$x_3 = -\frac{\det G_1}{\det G} = -\frac{3}{3} = -1.$$

It easily follows that $x_4 = -2$.

While certainly technical and concise, Dodgson's 1866 paper for the Royal Society would have been largely intelligible to the mathematically informed reader of the Society's *Proceedings*. However, its relatively straightforward and computational style stands in sharp contrast to the more abstract and rigidly formal presentation of his 1867 book.[31]

The *Elementary Treatise*

As we have seen, in his *Elementary Treatise on Determinants* Dodgson aimed to treat the subject by presenting a 'continuous chain' of reasoning, starting from fundamental definitions and proceeding via a sequence of rigorously proved propositions and theorems to a systematic treatment of the theoretical underpinning of the theory of determinants. The style is formal, axiomatic, and strictly logical, emulating the deductive structure of Euclid's *Elements*. In his opening preface he was quick to criticize the 'semi-colloquial semi-logical' styles of other mathematical authors and, in contrast to those modes of presentation, as a model of logical precision and exactitude his exposition is hard to fault. But it is similarly hard to read, as a passage from the book's third chapter illustrates:[32]

Corollary to Prop. X.

If there be n Equations, containing $\overline{n+r}$ Variables; and if there be among them $\overline{n-k}$ Equations, which have their V-Block not evanescent; and if, when these $\overline{n-k}$ Equations are taken along with each of the remaining Equations successively, each set of $\overline{n-k+1}$ Equations, so formed, has its B-Block evanescent (whence also $\|B\|=0$): the Equations are consistent; and, if any non-evanescent principal Minor of the V-Block of these $\overline{n-k}$ Equations be selected, the $\overline{k+r}$ Variables, whose coefficients are not contained in it, may have arbitrary values assigned to them; and, for each such set of arbitrary values, there is only one set of values for the other Variables; and the remaining Equations are dependent on these $\overline{n-k}$ Equations.

The general lack of readability caused by the book's rigid structure and awkward prose was further compounded by Dodgson's insistence on using his own self-devised notation and terminology, despite advice to the contrary from Spottiswoode. In particular, although well aware that the word *matrix* was already in wide circulation through the publications of Cayley and Sylvester, he argued:[33]

I am aware that the word 'Matrix' is already in use to express the very meaning for which I use the word 'Block'; but surely the former word means rather the mould, or form, into which algebraical quantities may be introduced, than an actual assemblage of such quantities; for instance, $\dfrac{(\quad)\times(\quad)}{(\quad)}$ would deserve the name, rather than $\dfrac{(a+b)\times(c+d)}{(e+f)}$.

He also used a rather cumbersome notation for the individual entries of a matrix. Whereas he had used the standard notation $a_{11}, a_{12}, a_{13}, \ldots$ in his earlier paper, in his book he adopted $1\iota1, 1\iota2, 1\iota3, \ldots$, since he objected to the superfluous use of the letter a and 'minute subscripts, alike difficult to the writer and the reader.'[34] While acknowledging that new words and symbols are never welcome in a subject with as large a technical lexicon as mathematics, he justified his introduction of them as 'the only way of avoiding tedious periphrasis'[35] – a dubious claim.

Dodgson's idiosyncratic style is symptomatic of his unorthodox approach to mathematics. In the words of one of his biographers:[36]

He read comparatively little of the works of other mathematicians or logicians, preferring to develop his theories out of his own mind

wherein any Element $\hbar\backslash \hbar_c = \{\hbar\backslash 1 \ldots\ldots \hbar\backslash r\}_a \S\{\hbar\backslash 1\ldots\ldots \hbar\backslash r\}_b$;

$$= \Sigma\{\hbar\backslash a_a . \hbar\backslash a_b\},$$

in which a takes all values from 1 to r;

now let a certain Constituent of the Determinant of the new Block be arranged in order of antecedents, and be represented by $1\backslash Q_a . 2\backslash R_a \ldots\ldots n\backslash T_a$, in which $Q, R, \ldots\ldots T$, are a certain permutation of the numbers 1 to n;

then this Constituent

$$= \Sigma\{1\backslash a_a . Q\backslash a_b\} . \Sigma\{2\backslash \beta_a . R\backslash \beta_b\}\ldots\ldots \Sigma\{n\backslash \delta_a . T\backslash \delta_b\};$$

in which *each* of the quantities $a, \beta, \ldots\ldots \delta$, takes all values from 1 to r;

$\therefore$ it $= \Sigma\{1\backslash a_a . Q\backslash a_b . 2\backslash \beta_a . R\backslash \beta_b \ldots\ldots n\backslash \delta_a . T\backslash \delta_b\}$;

$$= \Sigma\{1\backslash a_a . 2\backslash \beta_a \ldots\ldots n\backslash \delta_a . Q\backslash a_b . R\backslash \beta_b \ldots\ldots T\backslash \delta_b\};$$

also this Constituent is affected with $+$ or $-$, according as the series $Q, R, \ldots\ldots T$, contains an even or odd number of derangements;

$\therefore$ the Determinant of the new Block

$$= \Sigma\{1\backslash a_a . 2\backslash \beta_a \ldots\ldots n\backslash \delta_a . Q\backslash a_b . R\backslash \beta_b \ldots\ldots T\backslash \delta_b\},$$

in which not only does each of the quantities $a, \beta, \ldots\ldots \delta$, take all values from 1 to r, but also the series $Q, R, \ldots\ldots T$, takes the values of all possible permutations of the numbers 1 to n;

$\therefore$ it $= \Sigma\{1\backslash a_a . 2\backslash \beta_a \ldots\ldots n\backslash \delta_a . \Sigma(Q\backslash a_b . R\backslash \beta_b \ldots\ldots T\backslash \delta_b)\}$;

wherein, whatsoever values are assigned to $a, \beta, \ldots\ldots \delta$, in the outer bracket, the same are assigned to them in the inner bracket;

now the sum $\Sigma(Q\backslash a_b . R\backslash \beta_b \ldots\ldots T\backslash \delta_b)$, each term of which is affected with $+$ or $-$, according as the series $Q, R, \ldots\ldots T$, contains an even or odd number of derangements, is the Determinant

$$\begin{vmatrix} 1\backslash a \ldots\ldots 1\backslash \delta \\ \vdots \\ n\backslash a \ldots\ldots n\backslash \delta \end{vmatrix}_b ;$$

that is, it is the Determinant of the square Block formed by taking from the b-Block its ath column, its βth column, and so on, until n columns have been taken, it being immaterial whether these be all different, or one or more of them be repeated any number of times.

A page from the *Elementary Treatise*, illustrating Dodgson's bizarre notation

– an assessment borne out by the content of Dodgson's *Elementary Treatise*, much of which was original. Indeed, the only other textbook on determinants he cited was a German source, *Theorie und Anwendung der Determinanten* by Richard Baltzer, which he probably used in Jules Hoüel's French translation of 1861.[37] It was in this book that Dodgson would have come across Jacobi's theorem on determinants of adjugate minors,[38] as his lack of familiarity with mainstream mathematical research renders it unlikely that he would have read Jacobi's original papers. In any case, he never cited them, and he may even have been unaware that the result was due to Jacobi, since he never mentioned him by name.

It is also instructive to note that, although Dodgson's extensive private library housed books by mathematicians including George Boole, Augustus De Morgan, Benjamin and Charles Peirce, George Salmon, and Isaac Todhunter,[39] it contained no books on determinants – by Baltzer, Spottiswoode, or any other author. It is likely that any work he consulted on the subject was owned either by an Oxford library or by an academic colleague.

The rank of a matrix

In the 19th century many new concepts in algebra were in the process of formation, and it is not surprising that some of them emerged independently in the work of more than one mathematician. Among the many modern algebraic terms was the *rank* of a matrix, first formulated explicitly by the German mathematician Georg Frobenius in 1879.[40] In simple terms, the rank is a measure of how much the lines of a matrix depend on each other. For example, in both of the matrices

$$\begin{pmatrix} 3 & 2 & 1 \\ 1 & 0 & -1 \\ 2 & 2 & 2 \end{pmatrix} \quad \text{and} \quad \begin{pmatrix} 3 & 2 & 1 & 4 \\ 1 & 0 & -1 & -4 \\ 2 & 2 & 2 & 8 \end{pmatrix},$$

the first row is the sum of the second and third rows and so is a combination of the other two; thus, the rank of each matrix is 2. If the rows were completely independent of each other, the rank would be 3.

This has important implications for the study of systems of equations. For example, because these two matrices have the same rank, the system of equations associated with them,

$$\begin{aligned} 3x_1 + 2x_2 + x_3 &= 4 \\ x_1 - x_3 &= -4 \\ 2x_1 + 2x_2 + 2x_3 &= 8, \end{aligned}$$

must be solvable, or *consistent*. Indeed, this system has infinitely many sets of solutions, such as

$$x_1 = -1, x_2 = 2, x_3 = 3 \quad \text{and} \quad x_1 = 1, x_2 = -2, x_3 = 5.$$

On the other hand, if the second matrix were slightly different – that is, if the two matrices were

$$\begin{pmatrix} 3 & 2 & 1 \\ 1 & 0 & -1 \\ 2 & 2 & 2 \end{pmatrix} \quad \text{and} \quad \begin{pmatrix} 3 & 2 & 1 & 4 \\ 1 & 0 & -1 & -4 \\ 2 & 2 & 2 & 7 \end{pmatrix},$$

then the rank of the first matrix would still be 2, but the rank of the second would now be 3, since each row is no longer a combination of the other two. Because the ranks of the two matrices are no longer equal, their associated system of equations

$$3x_1 + 2x_2 + x_3 = 4$$
$$x_1 - x_3 = -4$$
$$2x_1 + 2x_2 + 2x_3 = 7$$

is *inconsistent*: no values of x_1, x_2, and x_3 satisfy all three equations simultaneously.

A system of linear equations is consistent (or solvable) if and only if the rank of its coefficient matrix is equal to the rank of its augmented matrix. This result is known today by a number of names, such as the *Kronecker–Capelli theorem*, the *Rouché–Capelli theorem*, and the *Rouché–Frobenius theorem*. This illustrates the occurrence of the same idea to different mathematicians in different contexts. Yet the first published proof of this theorem was actually due to Dodgson, as was noted in 1978 by two Russian historians of mathematics, Isabella Bashmakova and Aleksei Rudakov:[41]

The notion of rank of a matrix and the Kronecker-Capelli theorem were discovered independently by a number of investigators. The first published proof of this theorem is that of Charles L. Dodgson, author of the famous novels *Alice in Wonderland* and *Through the looking glass*. The theorem was published in his *An elementary treatise on determinants* in the following formulation:

For a system of n inhomogeneous equations in m unknowns to be consistent it is necessary and sufficient that the order of the largest nonzero minor be the same for the augmented and unaugmented matrices of the system.

This is misleading. Although Dodgson did indeed state and prove something close to this theorem, the above statement is not how he formulated it. His prose was more idiosyncratic:[42]

If there be given n Equations, not all homogeneous, containing Variables: a test for their being consistent is that either, first, there is one of them such that, when it is taken along with each of the remaining Equations successively, each pair of Equations, so formed, has its B-Block evanescent; or secondly, there are m of them, where m is one of the numbers 2. . . . n, which contain at least m Variables, and have their V-Block not evanescent, and are such that, when they are taken along with each of the remaining Equations successively, each set of Equations, so formed, has its B-Block evanescent.

This passage confirms why much of the content of Dodgson's *Treatise* went largely unnoticed by the mathematical community at large. As Francine Abeles has put it:[43]

Dodgson's convoluted style certainly reflects how his mind dealt with complicated problems. He often made connections that more conventional scholars might not see, particularly when subtle differences of meaning were involved. Unfortunately, his style also hindered broader acceptance of his stronger ideas like this one.

It also makes one appreciate how observant Bashmakova and Rudakov were to spot the equivalent formulation of a well-known theorem in such obscure and abstruse English!

Impact and influence

Looking back on Dodgson's *Treatise* half a century later, the Scottish mathematician Thomas Muir described it as 'a text-book quite unlike all its predecessors', both for its 'logical exactitude' and also for its 'use of more than the ordinary variety of printers' type'. But with Dodgson's extensive use of unorthodox notation, Muir noted that its pages presented 'an unwonted and rather bizarre appearance'.[44]

References to the book's unconventional symbols and terminology also appear in contemporary reviews, with the *Pall Mall Gazette* observing in 1868 that[45]

a student accustomed only to the old paths would not make much out of a dip into its pages.

Writing approvingly of 'the severely logical form into which it is thrown,' the anonymous reviewer gave the book a generally positive review, concluding that[46]

it serves to show the ability of Mr. Dodgson . . . and it creates a hope that we may have from him the benefit of further investigation in the algebraical wonderland of his choice.

Such sanguinity was not shared by the reviewer in the *Educational Times*, who held the opinion that it was Baltzer's – not Dodgson's – textbook on the subject that was 'a model of what such an elementary Treatise should be'.[47] Criticizing Dodgson's new notation for the entries of a matrix, he remarked that

the new symbol is not by any means an elegant one, nor do we think it is likely to find much favour with Mathematicians.

And although commending Dodgson's analysis of systems of equations, he commented somewhat disdainfully that his method of condensation[48]

does not seem on the whole to possess much advantage over the ordinary methods.

Dodgson did not seem too upset by this review, commenting that 'it is a more healthy one to read than if it were all praise', but he also observed:[49]

at the same time I cannot avoid a dim suspicion that the writer is not a mathematician and that he has not read much of the book.

For this reason, he was somewhat perplexed by the most critical remarks, contained in the review's final paragraph:[50]

In his desire to avoid what he calls the "semi-colloquial, semi-logical form often adopted by Mathematical writers," Mr. Dodgson runs into the opposite extreme, and makes such a parade of precision with "Definitions," "Conventions," "Axioms," "Proposition-Theorems," "Corollaries," "Enunciations," "Q.E.D's," &c., that the work becomes rather a tedious one to read. We should be glad to see much of this ostentation of formality excluded from subsequent editions of the work, if any such should be called for.

But this never happened: no further edition of Dodgson's *Treatise* ever appeared. Indeed, after he had completed his book at the end of 1867, Dodgson undertook no further work on the subject of determinants.

Both of these factors, taken with Dodgson's vagaries of style and notation, explain to a great extent why his work on determinants was largely ignored and had virtually no impact on the immediate subject or the larger study of algebra. Moreover, since university students for whom the book was mainly intended needed to know how to compute and manipulate determinants, rather than to reproduce logically precise proofs about them, the reviewer in the *Pall Mall Gazette* astutely predicted:[51]

It will be some time … before this book comes to be "got up" in the ordinary sense of the phrase by students; for we suspect it would scarcely "pay" in university examinations, and that, we understand, is the great final test which decides whether a subject shall be read exhaustively or not by our young mathematicians.

As a functional textbook, therefore, the work was essentially useless.

Throughout the 20th century Dodgson's work on determinants was largely ignored; as Thomas Muir wrote in 1920, 'it has certainly not commended itself to subsequent writers'.[52] His condensation method did feature in a few textbooks, however, including a numerical analysis text by P. S. Dwyer in 1951 and an algebra work by H. W. Turnbull in 1960.[53] But for the most part, Dodgson's contribution to algebra was mostly forgotten.

It was not until the mid-1980s that his method of condensation finally emerged from its long period of obscurity, when it was employed to create some completely new mathematics, as outlined in Chapter 7. This second, and far more fruitful, phase of its existence began in 1986 when David Robbins and Howard Rumsey[54] used an iteration of Dodgson's algorithm – referred to by Francine Abeles as 'Dodgson's determinantal identity',[55] or DDI – to formulate a hypothesis known as the *alternating sign matrix conjecture*, which quickly became an important unsolved problem in the mathematical area of combinatorics.[56] Its proof by Doron Zeilberger in 1995 prompted numerous further studies of determinant identities, using combinatorial and computational techniques. Since then, Dodgson condensation in the form of DDI has emerged as a remarkably fertile mathematical technique, serving in the derivation of a host of algebraic and combinatorial results.[57] Consequently, as the basis of an algorithm of tremendous utility in combinatorial and computational mathematics, the use of Dodgson's condensation method looks set to continue well into the 21st century.

𝔄 Syllogism worked out.

That story of yours, about your once meeting the sea=serpent, always sets me off yawning;
I never yawn, unless when I'm listening to something totally devoid of interest.

The Premisses, separately.

The Premisses, combined.

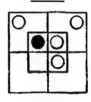

The Conclusion.

That story of yours, about your once meeting the sea=serpent, is totally devoid of interest.

Frontispiece of *Symbolic Logic Part I: Elementary*

CHAPTER 4

Logic

AMIROUCHE MOKTEFI

Charles Dodgson had a lifelong interest in logic. His firm belief in the utility of the subject in scientific practice, social affairs, and religious thinking led to work on a treatise that would simplify and popularize the subject, and for the purpose he introduced new diagrammatic and symbolic methods to solve logic problems. This practice earned him a place among the supporters of the new logic that was developing in the footsteps of the work of George Boole.

In this chapter we consider Dodgson's contribution to symbolic logic. After surveying the genesis of his interest in the subject we explore the general state of symbolic logic in his time, and in particular the types of problems that were addressed by symbolic logicians. Later sections are devoted to an exposition of Dodgson's methods to solve these problems. Finally, we address his work on hypotheticals and explain how it made him known to his contemporaries and successors. A further discussion of Dodgson's logic legacy appears in Chapter 7.

An obscure writer on logic

Although Dodgson's first publications on logic did not appear until the mid-1880s, his interest in the subject was several decades older. He was probably introduced to traditional logic in his early years as an undergraduate at Christ Church, and his diaries of 1855 already report activities in reading and writing logic.[1]

The Mathematical World of Charles L. Dodgson (Lewis Carroll). Robin Wilson and Amirouche Moktefi.
Oxford University Press (2019). © Oxford University Press 2019.
DOI: 10.1093/oso/9780198817000.001.0001

Dodgson's literary writings also contain numerous passages of substantial logical relevance.[2] However, it is in his mathematical investigations and his public pamphlets that we find the best illustrations of his concern for rigorous reasoning and exposition. In his *Elementary Treatise on Determinants* (see Chapter 3), he complained about 'the semi-colloquial semi-logical form often adopted by Mathematical writers'.[3] His championship of logic also motivated his opposition to the modern textbooks that were offered as substitutes to Euclid's *Elements* in geometrical teaching. In *Euclid and his Modern Rivals* (see Chapter 2), Dodgson declared that he expected a good manual of geometry to 'exercise the learner in habits of clear definite conception, and enable him to test the logical value of a scientific argument'.[4] Accordingly, he disqualified Euclid's rivals on the ground of their logical imprecision and their inferiority to Euclid's work, which should consequently be maintained in schools and colleges.

Dodgson also regularly intervened in public debates to discredit fallacious arguments. During his vacations at Eastbourne in the summer of 1877, for instance, he engaged in a dispute on the merits of vaccination to refute an argument that he held to be a fallacy. He declared that he 'did not come forward as a champion in the controversy, but as a critic; and [was] concerned rather with the logical accuracy of the weapons than with the result of the fight'.[5] Religious thinking also provided crucial motivation for Dodgson's work on logic.[6] Indeed, he believed that 'The bad logic that occurs in many and many a well-meant sermon, is a real danger to modern Christianity',[7] and even considered writing a book that would address religious difficulties treated 'from a logical point of view, in order to help those, who feel such difficulties, to get their ideas clear, and to see what are the logical results of the various views held'.[8]

It is not known when Dodgson became specifically interested in symbolic logic. An 1876 entry in his diaries shows that he was already familiar with Boole's algebraic notation,[9] but it was only from 1884 that logic, treated symbolically, appears as a regular preoccupation. On 29 March 1885 he included among his list of literary projects 'A symbolical Logic, treated by my algebraic method',[10] and in the following year he produced a little book entitled *The Game of Logic*, published under his pen-name of Lewis Carroll, in which players received instruction and amusement while solving syllogisms. As he recorded in his diary for 24 July 1886:[11]

The idea occurred to me this morning of beginning my "Logic" publication, not with "Book I" of the full work "Logic for Ladies," but with a small pamphlet and a cardboard diagram, to be called *The Game of Logic*. I have during the day written most of the pamphlet.

The *Game of Logic* first appeared in 1886 but, due to Dodgson's dissatisfaction with the quality of its printing, a new and revised edition was published in the following year.[12] It

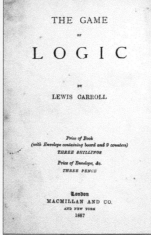
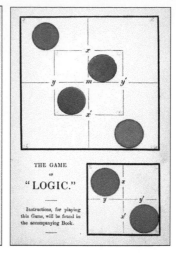

The Game of Logic, its title page, and the board with counters

was divided into four chapters – the rules of the game, a large collection of examples to be 'played', some solutions, and a set of unanswered examples – and was accompanied by a board on which two diagrams were printed, and some coloured counters.

Despite Dodgson's efforts to make *The Game of Logic* accessible, several reviewers failed to appreciate its combination of instruction and entertainment 'on the principle that two things, each good in itself, often make when mixed a third thing which is neither good nor desirable'.[13] Yet in the years that followed, Dodgson continued to use it in his logic classes, both private and public, in Oxford schools. He also continued to work on his treatise and in 1893 he revealed to his publisher his intention to have it published in three parts by level of difficulty: Elementary, Advanced, and Higher.[14]

Dodgson's long-awaited work *Symbolic Logic. Part I. Elementary* eventually appeared in February 1896. Two further editions appeared in the same year and a fourth edition was published in early 1897. The book first exposed Dodgson's theory of things, attributes, and propositions. He then introduced diagrammatic and symbolic methods to solve syllogisms and soriteses (we explain these terms below), and included a large collection of examples with answers and solutions. He took special care in the writing of his logic book to make it accessible to a wide audience and, like *The Game of Logic*, he signed the treatise with his literary pseudonym to give it extra publicity.

Prior to its publication Dodgson issued a pamphlet to promote his *Symbolic Logic*. He warned his readers against common misconceptions about logic being difficult, uninteresting, and useless, and then invited them to try his methods before making up their

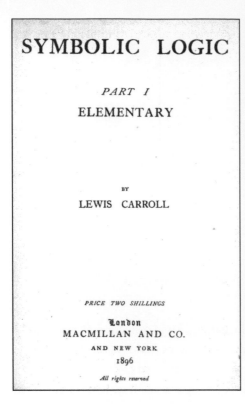

The original title page of *Symbolic Logic* and the Dover edition of 1958

minds.[15] He specifically praised the merits of symbolic logic in the way that it facilitates the solution of logic problems:[16]

Think of some complicated algebraical problem, which, if worked out with *x, y, z*, would require the construction of several intricate simultaneous equations, ending in an affected quadratic. Then imagine the misery of having to solve it in *words* only, and being forbidden the use of *symbols*. This will give you a very fair idea of the difference, in solving a Syllogism or Sorites, between the use of *Symbolic* Logic, and of *Formal* Logic as taught in the ordinary text-books.

Dodgson's book received few reviews and was seldom mentioned in the subsequent logic literature. It apparently did not catch the attention of logicians, partly because its content was elementary, as he had stated in the title of the book. Yet it was appreciated for its writing and for its large stock of examples which are still quoted nowadays in logic education. Hugh MacColl gave it a mixed review in *The Athenaeum*:[17]

It is well arranged, its expositions are lucid, it has an excellent stock of examples – many of them worked out, and not a few witty and amusing; and its arguments, even when wrong, are always acute and well worth weighing…On the whole, though the author has unquestionably written an interesting and useful little work, his proposed symbolic method does not appear to possess any advantages over some other methods that have preceded it.

Dodgson's death in 1898 prevented him from completing the remaining parts of his treatise. In the published Part I Dodgson had declared that Part II was supposed to cover 'all the ground usually traversed in the textbooks used in our Schools and Universities', while Part III was expected to 'deal with many curious and out-of-the way-subjects, some of which are not even alluded to in any of the treatises I have met with'.[18] Although most of this material is probably lost, if it had even been written, it is fortunate that some fragments survived and were published in 1977 by William Warren Bartley, III, who gathered galley proofs, manuscripts, notes, and various letters on logic. Although it can hardly be viewed as a definitive reconstruction of Dodgson's lost (and probably incomplete) work, Bartley offered a rich collection of previously unpublished logic material. His book was widely reviewed and certainly enhanced Dodgson's logical reputation as he made high claims for *Symbolic Logic* and argued that it revealed Dodgson as a highly innovative logician. Although Bartley's book was generally well received, several reviewers did not share his enthusiasm and accused him of exaggeration.[19]

Dodgson's lifetime and posthumous writings on logic offer a challenging set of material to the historian of logic. To make sense of his logical work one must keep in mind the important developments that the discipline of logic saw in his lifetime, especially in Britain: indeed, Dodgson's immediate predecessors and contemporaries greatly contributed to what is sometimes known as the 'mathematization' of logic. It is true that Dodgson was relatively isolated in Oxford and that he did not play an influential role in this development, describing himself modestly as 'an obscure Writer on Logic, towards the end of the Nineteenth Century'.[20] Yet he knew the ongoing work in symbolic logic,[21] and the title of his treatise alone suffices to attest to his belonging to the small but growing group of symbolic logicians.

The business of symbolic logic

Logic, as the science of correct reasoning, is naturally an ancient discipline. At the beginning of the 19th century it was mainly studied within the syllogistic tradition that originated with Aristotle. It is not true to claim that formal logic has made no significant progress since Aristotle, yet 'syllogistic' has remained the dominant doctrine and may roughly be characterized as follows.

A *syllogism* contains three propositions: two are its *premises* and the third is its *conclusion*, so that the premises necessarily yield the conclusion. For example, in the syllogism

All Greeks are Humans
All Humans are Mortals
Therefore, All Greeks are Mortals,

the first two statements are the premises and the last is the conclusion. In a syllogism each premise connects a common term ('Humans') with a distinct term ('Greeks' or 'Mortals'). The common term, called the *middle term*, is then eliminated in the conclusion which reveals the connection between the two distinct terms that survive.

Such propositions commonly indicate the predication of an attribute to a subject. This predication might be affirmed or denied, and may apply to the whole scope of the subject or to just part of it, and so logicians have commonly distinguished and named *four* canonical forms of propositions, depending on their quality and their quantity.

It is easy to recognize the structure of a given syllogism. Although there are infinitely many syllogisms, there are only finitely many possible structures: we have simply to enumerate the combinations of the types of proposition that form the syllogism and the position of the middle term in the premises. A simple calculation shows that there are 256 distinct structures, and it is then the business of a logician to identify and name the forms that are valid – that is, those in which the conclusion necessarily follows from the premises. For example, if M, S, and P are the given terms, then the following syllogistic form (1) is valid while the form (2) is not:

(1) No M are P; All S are M; therefore, No S are P.
(2) No M are P; No S are M; therefore, No S are P.

Among all the possible forms, logicians traditionally recognized twenty-four forms as valid. The problem that students of formal logic typically faced was to be given a syllogistic form and asked whether or not it was valid, and for the purpose one might simply compare it with the set of valid forms that had previously been learned by rote. Also, students learned some heuristic rules that simplified the solution of problems.

Logicians also addressed more complex problems, known as *soriteses*, involving more than three terms and two premises. However, such problems were commonly treated by reducing them to a series of syllogisms, again making the syllogism the main unit of reasoning. Symbolic and diagrammatic notations had also been progressively introduced, especially in the 18th-century German-speaking world, by such logicians as Gottfried Wilhelm Leibniz, Leonhard Euler, and Johann Heinrich Lambert, who provided quasi-mechanical methods of solution and opened the way to the development of symbolic logic.

At the beginning of the 19th century formal logic witnessed an important revival in Britain, notably through the work of Richard Whately, George Bentham, Sir William S. Hamilton, and Augustus De Morgan. In particular, under the influence of the developing symbolical algebra, De Morgan and George Boole together initiated a tradition of logic that made a thorough use of symbolism: Boole's 1854 *Investigation of the Laws of Thought* is commonly viewed as the main work in this direction.[22] Several logicians followed this path – notably, William Stanley Jevons, Alexander Macfarlane, and John Venn in Britain, Charles Sanders Peirce in the United States, Hugh MacColl in France, and Ernst Schröder in Germany.

Yet we should not necessarily think of symbolic logic as an immediate success story, because the new trend faced strong resistance from traditional logicians. We should rather think of a slow development in which the new ideas spread progressively. In particular, classification in the new logic challenged that of traditional logic, and opened the way to its modern association with mathematics, as argued by Boole:[23]

I am then compelled to assert, that according to this view of the nature of Philosophy, *Logic forms no part of it*. On the principle of a true classification, we ought no longer to associate Logic and Metaphysics, but Logic and mathematics…Logic resting like Geometry upon axiomatic truths, and its theorems constructed upon the general doctrine of symbols, which constitutes the foundation of the recognised Analysis.

Another characteristic of symbolic logic was its extensive appeal to symbolism, as suggested by its name. Although the expression itself was known prior to him, it was John Venn who popularized it through his book *Symbolic Logic*, first published in 1881. Venn defined the new logic as follows:[24]

In the first place, why do we entitle this subject Symbolic Logic? Is not all logic symbolic? Undoubtedly, it is, but the extent to which we employ symbols in one system and the other is so different that this really becomes one of the most obvious determining characteristics. The main distinction is this: that whereas the common logic uses symbols for classes, and for hardly anything else, we shall make equal use of symbols for operations upon these classes.

Finally, symbolic logicians also generalized and redefined the problems addressed within the discipline. In particular they worked extensively on the problem of elimination, which they viewed as the fundamental problem of logic. Jevons defined it as follows:[25]

George Boole…first put forth the problem of logical science in its complete generality: – *Given certain logical premises or conditions, to determine the description of any class of objects under those*

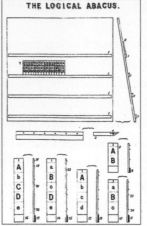

THE LOGICAL ABACUS.

W. Stanley Jevons, his 'logical abacus', and his logic machine

conditions. Such was the general problem of which the ancient logic had solved but a few isolated cases…Boole showed incontestably that it was possible, by the aid of a system of mathematical signs, to deduce the conclusions of all these ancient modes of reasoning, and an indefinite number of other conclusions. Any conclusion, in short, that it was possible to deduce from any set of premises or conditions, however numerous and complicated, could be calculated by his method.

The problem of elimination evidently appears as a generalization of earlier logic problems. No longer restricted to those involving just three terms, as demanded by syllogisms, symbolic logicians aimed at handling more complex problems involving any number of terms. Another important development in the problem of elimination was that logicians searched for the conclusion that follows from given premises, rather than checking the correctness of a conclusion that was given. Finally, they appealed to formal languages to handle this problem: many of them invented algebraic notations, others introduced diagrams, and a few (such as W. Stanley Jevons) even constructed logic machines, real and imaginary.

The general process for solving the elimination problem is roughly as follows, as shown in the following diagram: first 'translate' the concrete premises given in a natural language into a formal language; then process them to 'calculate' the conclusion that they entail; finally, 're-translate' the formal conclusion into the natural language to obtain the concrete conclusion of the problem.

Boole's followers engaged in a friendly competition to design the best notations and methods to solve the elimination problem. When Dodgson entered the scene, several systems were already in existence; for example, when Christine Ladd-Franklin introduced her algebra of logic in 1883, she declared that[26]

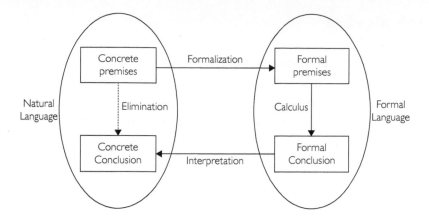

There are in existence five algebras of logic, – those of Boole, Jevons, Schröder, McColl, and Peirce, – of which the later ones are all modifications, more or less slight, of that of Boole. I propose to add one more to the number.

There were certainly several other algebras that Ladd-Franklin did not consider, and many more were to come. Dodgson added 'one more to the number'. We may even argue that he added more than one, because he introduced methods based on independent symbolic and diagrammatic notations. In later sections we briefly present Dodgson's notations and methods, but first, for the purpose of illustration and reference, we give a concrete example of a problem solved by earlier methods: Boole's algebraic notation and Venn's diagrams. Our example is an instance of the simplest of all syllogisms, known by traditional logicians as 'Barbara'.

The 'Barbara' problem

Suppose that we are given our two earlier propositions as premises:

P_1: All Greeks are Humans
P_2: All Humans are Mortals

and we wish to find the relation that connects Greeks and Mortals. To do so, we proceed as indicated in the elimination scheme, translating the premises into a formal language, calculating the conclusion, and then interpreting it. We will illustrate this process using Boole's algebra and Venn's diagrams.

Boole's algebra

Goerge Boole's algebraic notation was introduced in his 1847 *Mathematical Analysis of Logic* and expanded by him in his 1854 *The Laws of Thought*; subsequent notations by other logicians were mainly modifications or extensions of it. Noting a resemblance between the laws of logic and the laws of 'quantitative' algebra, Boole represented logical propositions as equations. Then the solution of logical problems merely requires the working-out of systems of equations, as mathematicians familiarly do.

George Boole and two pages from *The Laws of Thought*

To solve the Barbara problem above, we first need to rewrite the premises as equations. We note first that to assert that 'All Greeks are Humans' means that the intersection of the classes 'Greeks' and 'Humans' is the class 'Greeks' itself. Similarly, because all Humans are Mortals, the intersection of the classes 'Humans' and 'Mortals' is the class 'Humans'. So our two premises can be rephrased as:

P_1^*: The intersection of 'Greeks' and 'Humans' is 'Greeks'

P_2^*: The intersection of 'Humans' and 'Mortals' is 'Humans'.

Let us now substitute G, H, and M for 'Greeks', 'Humans', and 'Mortals'. Then these propositions P_1^* and P_2^* become:

P_1^{**}: The intersection of G and H is G

P_2^{**}: The intersection of H and M is H.

In Boole's notation the intersection of classes is represented as, and treated as analogous to, numerical multiplication, so our premises can be rewritten as equations:

$$E_1: G.H = G \quad \text{and} \quad E_2: H.M = H.$$

Replacing H in equation E_1 by its equivalent value $H.M$ found in equation E_2 then gives

$$E_3: G.(H.M) = G.$$

But intersection is associative, so this equation can be rewritten as

$$E_4: (G.H).M = G.$$

It now suffices to replace $G.H$ in E_4 by its equivalent value G found in E_1, to obtain the final equation:

$$E_5: G.M = G,$$

which tells us that:

C^{**}: The intersection of G and M is G.

If we now replace G and M by the terms that they stand for ('Greeks' and 'Mortals'), we obtain the interpretation:

C^*: The intersection of 'Greeks' and 'Mortals' is 'Greeks'.

This proposition, expressed in natural language, provides the desired conclusion:

C: All Greeks are Mortals.

Venn's diagrams

Venn first published his diagrams in the *Philosophical Magazine* in 1880, declaring that his diagrams 'underlie Boole's method' and are 'the appropriate diagrammatic representation of it'.[27] In the following year he presented at length his diagrams in his book *Symbolic Logic*; a second edition appeared in 1894.

SYMBOLIC LOGIC

BY

JOHN VENN, M.A.,

FELLOW, AND LECTURER IN THE MORAL SCIENCES,
GONVILLE AND CAIUS COLLEGE, CAMBRIDGE.

"Sunt qui mathematicum vigorem extra ipsas scientias, quas vulgo mathematicas appellamus, locum habere non putant. Sed illi ignorant, idem esse mathematice scribere quod in forma, ut logici vocant, ratiocinari."

LEIBNITZ, *De vera methodo Philosophiæ et Theologiæ* (about 1690).

"Cave ne tibi imponant mathematici logici, qui splendidas suas figuras et algebraicos mæandros universale inventionis veri medium crepant."

RÜDIGER, *De sensu veri et falsi*, Lib. II. Cap. IV. § xi. (1722).

𝕷onðon :

MACMILLAN AND CO.

1881

---

114 | *Diagrammatic Representation.* [CHAP.

shade out the compartments in our figure which have thus been successively declared empty, nothing is easier than to go on doing this till all the information furnished by the data is exhausted.

As another very simple illustration of the contrast between the two methods, consider the case of the disjunction, 'All *x* is either *y* or *z*'. It is very seldom even attempted to represent such propositions diagrammatically, (and then, so far as I have seen, only if the alternatives are mutually exclusive), but they are readily enough exhibited when we regard the one in question as merely extinguishing any *x* that is neither *y* nor *z*, thus :—

If to this were added the statement that 'none but the *x*'s are either *y* or *z*' we should meet the statement by the abolition of $\bar{x}y$ and $\bar{x}z$, and thus obtain :—

And if, again, we erase the central, or *xyz* compartment, we have then made our alternatives exclusive ; i.e., the *x*, and it alone, is either *y* or *z* only.

Now if we tried to do this by aid of Eulerian circles we should find at once that we could not do it in the only way in which intricate matters can generally be settled, viz., by

John Venn and two pages from his *Symbolic Logic*

Before solving the Barbara problem with Venn diagrams, we first note that to assert that 'All Greeks are Humans' simply means that there are no Greeks who are not Humans, so the intersection of the classes 'Greeks' and 'not-Humans' produces an empty class. Similarly, because all Humans are Mortals, the intersection of the classes 'Humans' and 'not-Mortals' also produces an empty class. So our two premises can be rephrased as:

P_1^*: The intersection of 'Greeks' and 'not-Humans' is the empty class
P_2^*: The intersection of 'Humans' and 'not-Mortals' is the empty class.

Let us now write G, H, and M for 'Greeks', 'Humans', and 'Mortals', and represent the empty class by '0'. Then these propositions become:

P_1^{**}: The intersection of G and not-H is 0
P_2^{**}: The intersection of H and not-M is 0.

Because the problem involves three terms G, H, and M, we need a Venn diagram for three terms. Each term is represented by a circle, and the three circles G, H, and M are drawn so as to depict the eight possible combinations of these terms.

Proposition P_1^{**} now asserts that the intersection of G and not-H is empty: we indicate this by shading the compartments common to G and not-H (that is, the space inside circle G and outside circle H).

Similarly, Proposition P_2^{**} asserts that the intersection of H and not-M is empty, so we additionally shade the compartments common to H and not-M (that is, the space inside circle H and outside circle M). The resulting diagram represents all the information contained in the premises.

To 'calculate' the relation between G and M, we focus on the relation between the circles G and M. Eliminating the circle H produces the following diagram.

We note that the space inside *G* and outside *M* is empty, which means that:

C**: The intersection of *G* and not-*M* is 0.

If we now replace *G*, *M*, and 0 by the terms 'Greeks', 'Mortals', and the empty class, we obtain the interpretation:

C*: The intersection of 'Greeks' and 'not-Mortals' is the empty class.

This proposition, expressed in natural language, provides the desired conclusion:

C: All Greeks are Mortals.

Solving the Barbara problem with the methods of Boole and Venn well illustrates the general solution process for the elimination problem. Although our example does not reflect symbolic logicians' efforts to tackle far more complex problems, it suffices for our purpose to observe how the problem was worked out within a formal language (be it algebraic or diagrammatic) in which 'calculations' were carried out. The early years of symbolic logic witnessed the invention of numerous notations and methods for this purpose, and in the following sections we present Dodgson's logic systems, his notations, and his methods.

The facts of logic and the facts of life

In Dodgson's logic the Universe contains objects that may or may not have certain given attributes (that is, qualities or properties). Classes are then formed by putting together certain objects. We may think, for example, of the class of 'Lions' by grouping all lions together, but in everyday life our discourses often have a narrower scope. Children visiting a zoo and expressing their desire to see 'the lions' are not referring to all existing lions, but only to the lions that are in the zoo. This implicit scope was named 'the universe of discourse' by 19th-century logicians,[28] and Dodgson recommended indicating explicitly the universe of each proposition or group of propositions.

Suppose that, within a determined universe of discourse, we form a class by grouping together the individuals that have a given attribute x. We have then instantly also formed another (negative) class containing those individuals that do not have that attribute x, and so may be said to have the attribute not-x. For instance, if the books on a given shelf form our universe of discourse, we can form the class of books that are green, and simultaneously also the class of books that are not green. This process of division by dichotomy produces distinct classes that together cover the entire universe of discourse.[29] Unlike many of his predecessors who had a '*morbid* dread of negative Attributes',[30] Dodgson regarded the complementary classes that are formed by dichotomy as being on the same footing:[31]

Under the influence of this unreasoning terror, they plead that, in Dichotomy by Contradiction, the *negative* part is too large to deal with, so that it is better to regard each Thing as either included in, or excluded from, the *positive* part. I see no force in this plea: and the facts often go the other way…For the purposes of Symbolic Logic, it is so *much* the most convenient plan to regard the two sub-divisions, produced by Dichotomy, on the *same* footing, and to say, of any Thing, either that it 'is' in the one, or that it 'is' in the other, that I do not think any Reader of this book is likely to demur to my adopting that course.

Dodgson distinguished two types of propositions, depending on what they asserted. A 'proposition of existence' asserts the status (existent or imaginary) of a given class, whereas a 'proposition of relation' asserts the relationship between classes, depending on the quantity (none, some, or all) of the members of a given class that belong to the other class:[32]

Propositions of existence: **E**. No x exist **I**. Some x exist

Propositions of relation: **E**. No x are y **I**. Some x are y **A**. All x are y

Interestingly, propositions of relation can be transformed into propositions of existence, as follows:

Propositions of relation		*Propositions of existence*
E. No x are y	→	No $x\,y$ exist
I. Some x are y	→	Some $x\,y$ exist
A. All x are y	→	No x not-y exist AND some $x\,y$ exist

Here the concatenation of two letters stands for the aggregation of the terms for which they stand; for example, within the universe of books, if x represents 'new' and y represents 'green', then $x\,y$ represents 'new green' books.

Most symbolic logicians viewed propositions of the form 'All x are y' as equivalent to 'No x are not-y'. Differing from this view, Dodgson defined propositions of the form 'All x are y' as double propositions that are equivalent to the conjunction of 'No x are not-y' and 'Some x are y';[33] for example, to assert that 'All animals are beautiful' is equivalent to the joint assertion that 'No animals are not-beautiful' and 'Some animals are beautiful'.

This definition has a decisive effect on the determination of what logicians call the 'existential import' of propositions.

Existential import refers to what a proposition asserts as to the existence (or non-existence) of its subject. It is commonly accepted that propositions of the form 'Some *x* are *y*' assert the existence of their subject; for example, when we declare that 'Some animals are beautiful', we assume that such animals exist. Contrariwise, propositions of the form 'No *x* are *y*' are commonly taken not to assert the existence for their subject. For example, when we affirm that 'No animals are not-beautiful' we do not assert (or deny) the existence of animals, but merely deny the existence of animals that would be not-beautiful.

Given that Dodgson defined propositions of the form 'All *x* are *y*' as containing those of the form 'Some *x* are *y*' which have existential import, it follows that propositions of the form 'All *x* are *y*' have an existential import as well.[34] For example, to state that 'All animals are beautiful' includes the assertion that 'Some animals are beautiful' which itself entails that 'Some animals exist'.

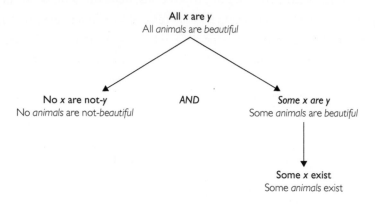

Unlike most of his colleagues in symbolic logic Dodgson refused to consider propositions of the form 'All *x* are *y*' as equivalent to 'No *x* are not-*y*', on the grounds that the former expresses the actual existence of the subject whereas the latter merely refers to it hypothetically:[35]

in every Proposition beginning with "some" or "all", the *actual existence* of the 'Subject' is asserted. If, for instance, I say "all misers are selfish," I mean that misers *actually exist*. If I wished to avoid making this assertion, and merely to state the *law* that miserliness necessarily involves selfishness, I should say "no misers are unselfish" which does not assert that any misers exist at all, but merely that, if any *did* exist, they *would* be selfish.

The attribution of an existential import to propositions starting with 'All' deeply complicated Dodgson's notations and methods, as we see in the following sections. His refusal to 'simplify' his system raises the important question of how truthful a formal language is expected to be when compared to natural languages. Venn openly favoured 'convenience and consistency in the working out of the Symbolic or Generalized Logic' and declared that he 'had to repudiate once for all any bounden obligation to either the language of common life, or that of the common logic' in the making of his symbolic logic.[36] Dodgson did not reject convenience and admitted that 'every writer may adopt his own rule, provided of course that it is consistent with itself and with the accepted facts of Logic'. However, when he came to decide about the existential import of propositions, he identified the views that 'may *logically* be held', tested them 'as applied to the actual *facts* of life', and rejected what he held to be 'singularly inconvenient for ordinary folk'.[37]

Dodgson worked on a logical theory that would be consistent with both the 'accepted facts of logic' and 'the actual facts of life'. This concern attested his firm belief in the social utility of symbolic logic and his confidence that his own treatise might prove useful to his readers to solve actual problems in ordinary life.[38] Logic was not to him a mere mental recreation; it was also a practical discipline. In the following sections, we present three methods designed by Dodgson to solve logical problems.

Diagrammatic methods

Dodgson first published his diagrams in 1886 in *The Game of Logic*, although an early version appears in his diary entry for 29 November 1884:[39]

Devised a way of working a syllogism, as opposite, Universe divided into eight categories, e.g. the upper right corner triangle is (gl'm'), i.e. "g, not-l, not-m," where "m" is the middle term, "g" the greater, i.e. the "major" term, and "l" the less, or minor.

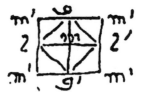

The extent of Dodgson's familiarity with the efforts in his time to design new diagrams among symbolic logicians is unclear. We may assume that he knew of Euler diagrams which were commonly used in traditional textbooks, but it is not known whether he was familiar with the attempts of Venn, Allan Marquand, and Macfarlane, who in the period 1880–81 produced schemes that resemble his.[40] Dodgson presented his diagrams at length in his 1897 treatise, but he never used them in print for problems other than syllogisms, even though he provided an interesting method of construction for more complex problems (see Chapter 7).[41]

To construct his diagrams Dodgson represented the universe of discourse as a square, which he then divided in order to represent terms. To represent a proposition of relation involving two terms such as *x* and *y*, he divided the square into four quarters correspond-ing to the combinations: *x y*, *x* not-*y*, not-*x y*, and not-*x* not-*y*, giving his 'biliteral diagram'. Note that the upper half of the square stands for *x*, the lower half for not-*x*, the left half for *y*, and the right half for not-*y*. Then, to represent propositions, he added marks to specify the status of the cells, with '0' indicating emptiness and '1' indicating existence.[42] For example, to represent the proposition 'No *x* are *y*', he put '0' in the cell *x y*, and to represent the proposition 'Some *x* are *y*' he put '1' in the cell *x y*.

x *y*	*x* not-*y*	0			I	
not-*x* *y*	not-*x* not-*y*					

In order to add a third term such as *m* to the previous diagram, he added a small square whose inside region stood for *m* and whose outside region stood for not-*m*, giving his 'triliteral diagram'. To represent propositions on this diagram he followed the same conventions as above, with '0' in a cell to indicate emptiness and '1' to depict existence.

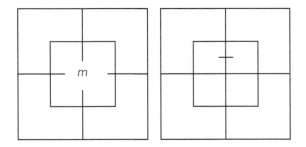

When it was known that at least one of two adjacent cells was occupied, but without further precision, Dodgson placed '1' on the boundary between the two cells as though it were 'sitting on the fence'.[43] For example, the proposition 'Some *x* are *m*' indicates that the region *x m* (upper half, inside the square) is occupied, and so should be marked with '1'. However, that region contains two cells (left and right) and it is not known which one (at least) is occupied, so we simply put '1' on the boundary between the two cells inside the targeted region *x m*, as shown above.

Biliteral and triliteral diagrams are used to solve logic problems, such as syllogisms, which have just three terms. We first represent the premises on the triliteral diagram, and then calculate the conclusion by eliminating the middle term and transferring information from the triliteral to the biliteral diagram from which the conclusion can be 'read off'. To operate the transfer we use the following two rules:[44]

Rule A: If a quarter of the triliteral diagram has 'I' in *either* cell, then we put 'I' in the corresponding quarter of the biliteral diagram.

Rule B: If a quarter of the triliteral diagram has two '0's, one in each cell, then we put '0' in the corresponding quarter of the biliteral diagram.

To illustrate the working of these rules, let us return to the propositions given as premises in the Barbara problem discussed earlier:

P_1: All Greeks are Humans
P_2: All Humans are Mortals

Within the Universe of creatures, let *h* stand for the middle term 'Humans', and let *g* stand for 'Greeks' and *m* for 'Mortals'. We then obtain the following premises in abstract form:

P_1^*: All *g* are *h*
P_2^*: All *h* are *m*

To find the conclusion that follows from these premises, we represent them on the triliteral diagram whose upper half stands for *g*, whose left half is *m*, and whose inside square is *h*.

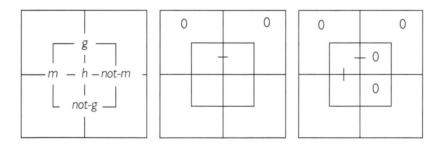

In accordance with Dodgson's definitions the first premise, 'All *g* are *h*', is a conjunction of two sub-propositions, 'No *g* are not-*h*' and 'Some *g* are *h*', and both of these must be represented on the diagram. The first asserts that 'No *g* not-*h* exist', which we depict by writing '0' in each cell of the region '*g* not-*h*'. The second asserts that 'Some *g h* exist', which we depict by writing 'I' in the region '*g h*' on the fence between its cells. The second premise 'All *h* are *m*' is similarly conceived as a double proposition, formed from the sub-propositions 'No *h* are not-*m*' and 'Some *h* are *m*'. We add these to the triliteral diagram, and then note that the region '*g h*' (upper half, inside square) is occupied but its right-hand cell '*g h* not-*m*' is empty, so that the left-hand cell '*g h m*' must be occupied. We can therefore move the '1' from the vertical fence into the region '*g h m*' and remove '1' from the horizontal fence as it has now become superfluous. We obtain a simplified triliteral diagram:

In order to find the conclusion, we must transfer the information in the triliteral diagram to a biliteral diagram that exhibits the relationship between the terms *g* and *m* (by eliminating *h*). To do so, we visit each quarter of the triliteral diagram and see what action is required by Dodgson's transfer rules. The upper-left quarter has '1' in one of its cells, so we apply Rule A and write '1' in the upper-left quarter of the biliteral diagram. The upper-right quarter has '0' in each cell, so we apply Rule B and write '0' in the upper-right quarter of the biliteral diagram. The lower quarters of the triliteral diagram entail no action to accord with Dodgson's rules. So we obtain the required biliteral diagram.

This expresses the conclusion of the syllogism and shows that 'Some *g* are *m*' and 'No *g* are not-*m*'. Combining these propositions, as explained earlier, yields

C*: All *g* are *m*

and when expressed in natural language, this gives us the desired conclusion,

C: All Greeks are Mortals.

Symbolic methods

In line with many of his British predecessors and contemporaries, Dodgson worked on several symbolic methods. As we have seen, he was already aware of Boole's algebraic notation by 1876, and a few years later his *Euclid and his Modern Rivals* of 1879 introduced several symbols that would be found in his later logic writings. But as his diary for

1884–86, suggests, it was during this period that he designed the subscript notation he would adopt and later publish in his 1896 *Symbolic Logic* Part 1.

In Dodgson's notation the terms are commonly represented by lower-case letters, such as *x*, *y*, *z*, etc. To indicate the negation of a term Dodgson added an accent mark to the letter representing the term: for example, if *x* stands for the term 'blue', then *x'* stands for 'not-blue'. To represent propositions of existence he introduced subscripts that are indexed to the symbol of a class to represent its state, with '0' indicating emptiness and '1' depicting existence; for example, if *x* stands for the term 'dog', then x_0 indicates that there are no dogs; such propositions asserting emptiness are called *nullities*. Similarly, x_1 indicates that there is at least one dog; such propositions asserting existence are called *entities*.

This method also allowed Dodgson to represent propositions of relation by converting them into their equivalent forms of existence, as described above. For example, to represent the proposition 'No *x* are *y*', we use its equivalent form 'No *x y* exist' and write it as xy_0, and similarly, to represent the proposition 'Some *x* are *y*', we use its equivalent form 'Some *x y* exist' and write it as xy_1. Finally, a proposition of the form 'All *x* are *y*' is equivalent to the conjunction of the propositions 'Some *x* exist' (written as x_1) and 'No *x* not-*y* exist' (written as xy'_0), and so we can write it as $x_1y'_0$, always bearing in mind that '*each* Subscript takes effect back to the *beginning* of the expression'.[45] The following list summarizes the symbolic expressions for the common propositions of existence and relation:

$$x_1 : \quad \text{Some } x \text{ exist}$$
$$x_0 : \quad \text{No } x \text{ exist}$$
$$xy_1 : \quad \text{Some } x \text{ are } y$$
$$xy_0 : \quad \text{No } x \text{ are } y$$
$$x_1y'_0 : \quad \text{All } x \text{ are } y$$

In addition to the previous symbols that represent the states and relations of classes, Dodgson introduced several more symbols for connectives between propositions:

$$P' \qquad : \quad \text{not-P}$$
$$P \dagger Q \quad : \quad \text{P and Q}$$
$$P \S Q \quad : \quad \text{P or Q}$$
$$P \, \mathbb{P} \, Q \quad : \quad \text{P implies Q}$$
$$P \equiv Q \quad : \quad \text{P is equivalent to Q}$$

Two of these symbols are particularly useful when it comes to expressing syllogisms: the dagger sign '†' stands for conjunction, and the sign '℗' means 'would, if true, prove'.[46] To

Pairs of Premisses for Syllogisms. *(Answers only are supplied.)*

1. All pigs are fat;
   Nothing that is fed on barley-water is fat.
2. All rabbits, that are not greedy, are black;
   No old rabbits are free from greediness.
3. Some pictures are not first attempts;
   No first attempts are really good.
4. Toothache is never pleasant;
   Warmth is never unpleasant.
5. I never neglect important business;
   Your business is unimportant.
6. No pokers are soft;
   All pillows are soft.
7. Some lessons are difficult;
   What is difficult needs attention.
8. All clever people are popular;
   All obliging people are popular.
9. Thoughtless people do mischief;
   No thoughtful person forgets a promise.
10. Pigs cannot fly;
    Pigs are greedy.

Dodgson's use of notation to represent syllogisms

represent a syllogism with this notation Dodgson wrote the three propositions in a row (each expressed with the subscript method), with the symbol '†' between the premises and '℗' before the conclusion.

In order to solve syllogisms, Dodgson distinguished just three figures for all syllogisms. It was then sufficient for him to determine the figure to which a pair of premises belonged in order to state what conclusion it entailed. Dodgson was proud of his method and claimed its superiority over traditional methods:[47]

As to *Syllogisms*, I find that [in ordinary text-books] their nineteen forms, with about a score of others which they have ignored, can all be arranged under *three* forms, each with a very simple Rule of its own; and the only question the Reader has to settle, in working any one of the 101 Examples…of this book, is "Does it belong to Fig. I. II., or III.?"

To determine to which of the three figures of syllogisms a given form belongs, Dodgson provided the algorithmic method shown opposite;[48] here the *retinends* are the terms that remain in the conclusion, and the *eliminands* are the middle terms that are excluded. Two eliminands are said to be 'like' if they have the same sign.

To illustrate the use of this algorithm, let us again consider the Barbara problem with premises:

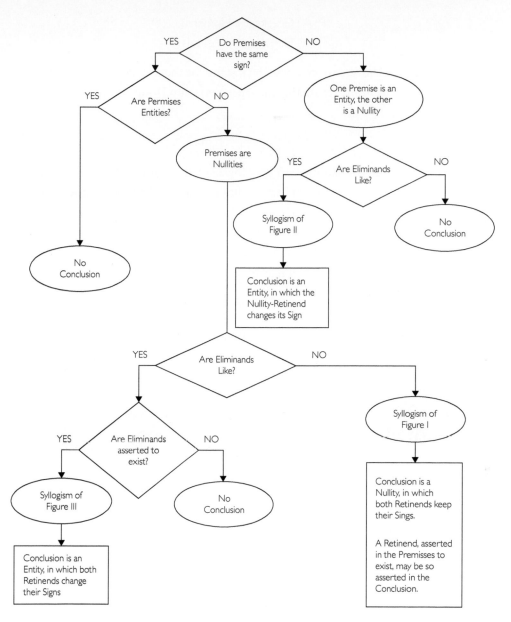

Dodgson's algorithm for solving syllogisms

P$_1$: All Greeks are Humans
P$_2$: All Humans are Mortals.

As usual, within the Universe of creatures, let h stand for the middle term 'Humans', g for 'Greeks', and m for 'Mortals'. Translating these premises into Dodgson's notation gives the symbolic forms:

$$P_1^*: g_1 h'_0 \qquad P_2^*: h_1 m'_0.$$

To reach the desired conclusion we follow the algorithm for our pair of premises. We observe that they have the same sign but are not entities, and so are nullities. We then note that the middle term h is denied in the first premise and affirmed in the second, and so the eliminands are not like. It follows that we have a syllogism whose conclusion is a nullity, in which both retinends (g and m) keep their signs, gm'_0. Because the term g is asserted to exist in the first premise, it may be so asserted in the final conclusion:

$$C^*: g_1 m'_0.$$

This proposition, expressed in natural language, provides the desired conclusion:

C: All Greeks are Mortals.

The method of trees

Dodgson invented his method of trees on 16 July 1894, as he recorded in his journal:[49]

Today has proved to be an epoch in my Logical work. It occurred to me to try a complex Sorites by the method I have been using for ascertaining what cells, if any, survive for possible occupation when certain nullities are given. I took one of the 40 premisses, with "pairs [...] within pairs," and many bars, and worked it like a genealogy, each term proving all its descendants. It came out beautifully, and much shorter than the method I have used hitherto. I think of calling it the "Genealogical Method."

Unlike the previous diagrammatic and symbolic methods presented earlier in this chapter, the method of trees was not included in the first part of Dodgson's *Symbolic Logic* which was published in his lifetime, as he evidently intended to include it in the later parts. On 6 November 1896 he completed a chapter devoted to this method and sent it for opinion to John Cook Wilson, the Oxford Professor of Logic. Fortunately Cook Wilson never returned the proofs, which were subsequently found and published by Bartley in his 1977 edition of Dodgson's surviving logic fragments.

John Cook Wilson   Christine Ladd-Franklin

The method of trees might be said to combine diagrammatic and symbolic features. It has an interesting procedure which consists in denying an expected conclusion that would follow from a given set of premises, and then searching for inconsistency. Dodgson described his method as follows:[50]

Its essential feature is that it involves a *Reductio ad Absurdum*. That is, we begin by assuming, *argumenti gratia*, that the aggregate of the Retinends (which we wish to prove to be a *Nullity*) is an *Entity*: from this assumption we deduce a certain result: this result we show to be *absurd*: and hence we infer that our original assumption was *false*, i.e. that the aggregate of the Retinends is a *Nullity*.

Francine Abeles has observed that this method resembles that of the Antilogism by Christine Ladd-Franklin, first published in 1883 in the volume of *Studies in Logic* that included works by C. S. Peirce and his students from Johns Hopkins University.[51] It is known that Dodgson owned a copy of this book.

In order to illustrate the working of the method of trees, let us return again to our Barbara problem, with the following premises:

P_1: All Greeks are Humans
P_2: All Humans are Mortals.

We maintain the symbols h for 'Humans', g for 'Greeks', and m for 'Mortals'. As before, translating these premises into Dodgson's notation gives the symbolic forms:

$$P_1^*: g_1 h'_0 \qquad P_2^*: h_1 m'_0.$$

Because h is the middle term we expect the conclusion to assert the relation between the other terms g and m'. Suppose that the conclusion is an entity, which means that there exist things that have both attributes g and m'. We then take gm' as the root of our tree:

$$\boxed{g\,m'}$$

The first premise asserts that g and h' are incompatible, so if a thing has the attribute g, then it must also have the attribute h (which is the negative of h'). We place h below the root gm' in the tree, with reference number '1' to keep track of the premise that gave us this information. The tree becomes:

$$\boxed{\begin{array}{l} g\,m' \\ 1.\,h \end{array}}$$

From the second premise it follows that h and m' are incompatible, so if a thing has the attribute m', then it must also have the attribute h' (which is the negative of h). So we add h' under the root $g\,m'$ in the tree, with reference number '2'. The tree becomes:

$$\boxed{\begin{array}{l} g\,m' \\ 2,1.h\,h' \end{array}}$$

But anything that has the two attributes h and h' is a nullity, since h and h' are incompatible. To indicate this we place a mark ○ under $h\,h'$ in the tree. We obtain:

$$\boxed{\begin{array}{c} g\,m' \\ 2,1.h\,h' \\ \circ \end{array}}$$

The tree indicates that anything with the attributes g and m' must also have the attributes h and h'. But anything with the attributes h and h' is a nullity. So our initial assumption that there exist things with both the attributes g and m' was incorrect. It follows that $g\,m'$ is a nullity. This result is represented in the tree as follows:

$$\boxed{\begin{array}{c} g\,m' \\ 2,1.h\,h' \\ \circ \\ g\,m'\!_\circ \end{array}}$$

We now examine whether g or m' is asserted to exist in the premises, and observe that g exists in the first premise. Introducing this information we obtain the complete tree:

$$\boxed{\begin{array}{c} g\,m' \\ 2,1.h\,h' \\ \circ \\ g_1 m'\!_\circ \end{array}}$$

This tree shows that the final conclusion is

$$C^{**}: g_1 m'_0.$$

This conclusion is the symbolic form of the abstract conclusion

$$C^*: \text{All } g \text{ are } m,$$

which, expressed in natural language, provides the desired conclusion:

$$C: \text{All Greeks are Mortals.}$$

We applied the method of trees to this simple example to exhibit the principle and the working of the method, but its real force is better revealed when applied to more complex problems. Dodgson designed numerous sorites problems, sometimes with dozens of premises, and also introduced generalizations and extensions of his methods to handle these problems, both symbolically and with diagrams.[52] Although we cannot address them here, one might consider some of the examples that Dodgson included in his treatise in order to realize the difficulty of handling such intricate problems.

Dodgson's methods have many merits (see Chapter 7). His diagrams have a closed Universe of discourse, a device for the representation of non-emptiness, explicit rules for the derivation of the conclusion, and an easy construction procedure for higher numbers of terms. His symbolic method also has several advantages: it offers a convenient scheme to represent existential statements, which troubled his British colleagues, and it clearly distinguishes notations for classes and propositions. Finally, the method of trees introduces an original decision procedure that would later be further explored by 20th-century logicians working with tableaux methods.

Yet, Dodgson's methods received little attention in his lifetime and hardly influenced his immediate followers. It is rather Dodgson's work on hypotheticals that introduced him to the community of logicians and secured his reputation among them.

The puzzling subject of hypotheticals

Although Dodgson's interest in hypotheticals can be traced in his earlier writings, his investigation of the subject seems to have intensified in the early 1890s.[53] In particular, he exchanged views on the subject with Cook Wilson in the period 1892–94. This work led to two papers in the philosophy journal *Mind*: 'A logical paradox' in 1894 and 'What the Tortoise Said to Achilles' in 1895. The former article made Dodgson better known to contemporary logicians, whereas the latter has been thoroughly

Bertrand Russell

debated by Dodgson's successors and is still widely discussed by modern philosophers and logicians.

Hypotheticals (also called *conditionals*) are statements of the form: 'If P, then Q', and are generally understood to assert that whenever P occurs, then Q must also occur. The relationship between the terms of the hypothetical (usually referred to in modern logic as the *antecedent* and the *consequent*) has been a major topic of discussion among logicians. Dodgson's articles on the subject are often considered among his best logical contributions; for example, Bertrand Russell discussed both articles in his 1903 book on *The Principles of Mathematics*, and later declared:[54]

I think [Dodgson] was very good at inventing puzzles in pure logic. When he was quite an old man, he invented two puzzles he published in a learned periodical, *Mind*, to which he didn't provide answers. And the providing of answers was a job, at least so I found it…His works were just what you would expect: comparatively good at producing puzzles and very ingenious and rather pleasant, but not important…None of his works was important. The best work he ever did in that line was the two puzzles that I spoke of.

The barber-shop problem

Dodgson's first contribution, 'A logical paradox', appeared in July 1894, and reported on a dispute between himself and Cook Wilson on the nature of hypotheticals. Their argument apparently started at the end of 1892 and can be tracked through Dodgson's entries in his journal; for example, on 5 February 1893 he recorded:[55]

Heard from Cook Wilson, who has long declined to read a paper, which I sent January 12, and which seems to me to *prove* the fallacy of a view of his about Hypotheticals.

A year later, on 1 February 1894, Dodgson reported again on the dispute:[56]

I got, from Cook Wilson, what I have been so long trying for, an *accepted* transcript of the fallacious argument over which we have had an (apparently) endless fight. I think the end is near, *now*.

A LOGICAL PUZZLE.

There are three Propositions, A, B, and C.
It is given that
"If A is true, B is true ; ....(i)
 If C is true, then if A is true B is not true" ......(ii)

Nemo and Outis differ about the truth of C.
Nemo says C cannot be true : Outis says it can

Nemo's Argument.

Number (ii) amounts to this :—
 "If C is true, then (i) is not true".
But, *ex hypothesi*, (i) *is* true.
 ∴ C cannot be true ; for the assumption of C involves an absurdity.

Outis's Reply.

Nemo's two assertions, "if C is true, then (i) is not true" and "the assumption of C involves an absurdity", are erroneous.

The assumption of C *alone* does *not* involve any absurdity, since the two Hypotheticals, "if A is true B is true" and "if A is true B is not true", are *compatible*; i.e. they can be true together, in which case A cannot be true.

But the assumption of C and A *together does* involve an absurdity; since the two Propositions, "B is true" and "B is not true", are *incompatible*.

Hence it follows, not that C, *taken by itself*, cannot be true, but that C and A cannot be true *together*.

Nemo's Rejoinder.

Outis has wrongly divided Protasis and Apodosis in (ii).

The absurdity is not the last clause of (ii), "B is not true", but *all* that follows the word "then", i.e. the Hypothetical "If A is true B is not true"; and, by (ii), it is the assumption of C only which causes this absurdity.

In fact, Outis has made (ii) equivalent to "If C is true [and if A is true] then if A is true B is not true". This is erroneous : the words in the brackets in the compound Protasis are superfluous, and the remainder is the true Protasis which conditions the absurd Apodosis, as is evident from the form of (ii) originally given.

The dispute between Dodgson (Outis) and John Cook Wilson (Nemo)
on the barber-shop paradox spilled over into print

Contrary to Dodgson's expectations, the fight had merely started. Indeed, the failure of Dodgson and Cook Wilson to agree on the outcome of their dispute made them solicit the opinion of their colleagues, with the result that the argument between the two men turned into a huge controversy that for a decade involved Britain's leading logicians, such as John Venn, Francis H. Bradley, Henry Sidgwick, John N. Keynes, William E. Johnson, Alfred Sidgwick, Hugh MacColl, E. E. Constance Jones, and Bertrand Russell.[57] Dodgson declared that the contradictory opinions he had privately collected motivated him to write his article:[58]

The paradox, of which the foregoing paper is an ornamental presentment, is, I have reason to believe, a very real difficulty in the Theory of Hypotheticals. The disputed point has been for some time under discussion by several practised logicians, to whom I have submitted it; and the various and conflicting opinions, which my correspondence with them has elicited, convince me that the subject needs further consideration, in order that logical teachers and writers may come to some agreement as to what Hypotheticals *are*, and how they ought to be treated.

Dodgson's problem, commonly known as *the barber-shop problem*, is the following.

Allen, Brown, and Carr are three barbers working in a shop under the following rules:
    1: If Carr is out, then (if Allen is out, then Brown must be in).
    2: If Allen is out, then Brown must be out.
Can Carr leave the shop?

Answering this looks rather easy: we simply construct a table showing the eight possible cases, depending on the presence or absence of each barber:

	1	2	3	4	5	6	7	8
Allen:	out	out	out	out	in	in	in	in
Brown:	out	out	in	in	out	out	in	in
Carr:	out	in	out	in	out	in	out	in

Now Rule 1 forbids the first case and Rule 2 excludes cases 3 and 4, so there are five remaining cases that are compatible with the rules. Among these, cases 5 and 7 show that Carr can go out of the shop without violating the rules. The issue seems settled.

However, Dodgson invites us to consider (but not necessarily approve) a different argument that can be outlined as follows. Let us make the assumption that Carr goes out of the shop. It follows from the first rule that:

    3. If Allen is out, then Brown must be in.

But this is said to be incompatible with Rule 2, so our initial assumption was wrong and Carr cannot go out, contradicting our previous solution.

The fallacy lies in the fact that the propositions

2. If Allen is out, then Brown must be out.
3. If Allen is out, then Brown must be in.

are actually not incompatible. They may very well both be true, in which case we can infer that Allen is in the shop. This does not conflict with our assumption of Carr being out of the shop. As Russell later wrote:[59]

The principle that false propositions imply all propositions solves Lewis Carroll's logical paradox.

What made Dodgson's argument intriguing to his contemporaries is that it offered a powerful illustration of what later became known as the 'paradoxes of material implication'.

What the Tortoise Said to Achilles

Unlike the barber-shop problem, which raised a large controversy, Dodgson's second contribution to *Mind* was not commented on during his lifetime. However, it later became one of the classic philosophy problems of the 20th century.

'What the Tortoise Said to Achilles' appeared in the journal *Mind* in April 1895. Although little is known about its genesis, it evidently sprang from Dodgson's work on the subject in 1894, and may have benefited from his exchanges with other logicians (notably, Cook Wilson) in that year. In his article Dodgson exposed a difficulty that has been commonly known since then as the *paradox of inference*.

Suppose we are given the following inference:

A: Things that are equal to the same are equal to each other.
B: The two sides of this Triangle are things that are equal to the same.

Therefore,

Z: The two sides of this Triangle are equal to each other.

It is said that a reader who accepts the premises A and B cannot be 'logically' forced to accept Z if he did not accept beforehand that:

C: If *A* and *B* are true, then *Z* must be true.

We might think that a reader who accepts A, B, and C is obliged to accept Z. However, Dodgson's argument suggests that such a reader still needs to accept beforehand that

D: If *A* and *B* and *C* are true, then *Z* must be true.

And so on. The necessity of further hypotheticals can be claimed at each stage, ad infinitum.

This infinite regress can be illustrated as follows. At each stage a new hypothetical, stating the validity of the inference, is added as a premise to the inference itself, thereby producing a new inference with an extra premise:

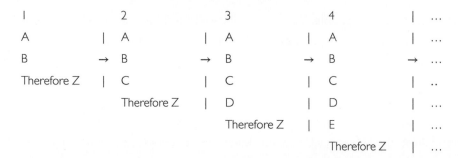

Dodgson's argument suggests that no solid inference can ever be made because the reasoner is trapped in an infinite regress. The problem has been thoroughly discussed by philosophers and logicians who attempted to stop the regress. An early comment was offered by Russell, who argued that Dodgson's story illustrates the need for a principle according to which[60]

A true hypothesis in an implication may be dropped, and the consequent asserted.

Although there is still no agreement as to the moral of the story, philosophers commonly contend that an inference should not include its own hypothetical. This view has been thoroughly defended by Gilbert Ryle who argued that[61]

The principle of an inference cannot be one of its premises or part of its premises. Conclusions are drawn from premises in accordance with principles, not from premises that embody those principles. The rules of evidence do not have to be testified to by the witnesses.

In his articles in *Mind* Dodgson did not openly express his views on hypotheticals. Rather, he presented two difficulties that logicians face when they investigate the subject. His journal shows that after writing these articles he continued to explore the issue in search for a 'workable theory of Hypotheticals'.[62] The subject is also listed among those to be covered in the lost parts of Dodgson's treatise. Although it is not known what the ultimate outcome of Dodgson's work on hypotheticals would have been, his two articles attest his interest and understanding of the complexity of the subject.

The problems that Dodgson raised clearly relate to the material interpretation of implication, but Bartley also perceived in them a 'foretaste' of strict and causal implication.[63]

More than his methods to solve syllogisms and soriteses, Dodgson's work on hypotheticals introduced him to his logical colleagues and secured his logical reputation among modern logicians.

Conclusion

Dodgson was a symbolic logician. Like few other Victorian logicians he designed symbolic methods to solve the elimination problem of finding what conclusion follows from a set of propositions offered as premises. Although his standing within the growing symbolic logic community is disputed, there is no doubt that he promoted the new logic. In Oxford he faced the opposition of Cook Wilson, the Professor of Logic, who had little sympathy for symbolic logic and believed that[64]

the real and serious problems of logic proper do not appear, nor is the symbolic logician able to touch them. In comparison with the serious business of logic proper, the occupations of the symbolic logician are merely trivial.

Dodgson's work may be viewed as a step towards the recognition of symbolic logic as a discipline with a 'serious business'. He was certainly proud of his symbolic methods and believed that they overcame the traditional procedures found in ordinary formal logic textbooks. In a letter to his publisher, dated 19 October 1895, he expressed his conviction in the superiority of symbolic logic:[65]

I have no doubt that Symbolic Logic (not necessarily *my* particular method, but *some* such method) will, *some* day, supersede Formal Logic, as it is immensely superior to it: but there are no signs, as yet, of such a revolution.

Although symbolic logic lost numerous battles in Dodgson's time, it eventually won the war and Dodgson's prediction proved true.

LAWN TENNIS

TOURNAMENTS

THE TRUE METHOD OF ASSIGNING PRIZES WITH A PROOF OF THE FALLACY OF THE PRESENT METHOD

BY

CHARLES L. DODGSON, M.A.

STUDENT AND LATE MATHEMATICAL LECTURER OF
CH. CH. OXFORD

𝔓𝔞𝔩𝔪𝔞𝔪 𝔮𝔲𝔦 𝔪𝔢𝔯𝔲𝔦𝔱 𝔣𝔢𝔯𝔞𝔱

PRICE SIXPENCE

LONDON

MACMILLAN AND CO.

1883

The title page of *Lawn Tennis Tournaments*, 1883

CHAPTER 5

Voting

IAIN McLEAN

Voting is not as simple as it may seem. For example, in a tournament with at least three candidates *A*, *B*, and *C*, and at least three voters, *A* can win a majority against *B*, who in turn wins a majority against *C*, who wins a majority against *A*. This situation was pointed out by the Marquis de Condorcet in 1785,[1] but his work was lost for almost a century. In 1876 it was given its modern name of a 'cycle' by C. L. Dodgson.[2] Dodgson's work on the mathematics of voting then fell into neglect for a century until it was rediscovered by Duncan Black and other writers.

Voting theory before Dodgson

Dodgson is one of the great figures in the axiomatic theory of voting, which is the study of cycling and other paradoxes and the properties of choice systems. His only peers are Ramon Llull (*c*.1235–1315), Nicholas of Cusa (1401–64), the Marquis de Condorcet (1743–94), Jean-Charles de Borda (1733–99), and (more recently) Duncan Black (1908–91) and Kenneth J. Arrow (1920–2017).[3]

Several of these thinkers were, like Dodgson, strikingly eccentric. All, like Dodgson, saw that the simplest version of majority rule is seriously flawed. This version is the one used in parliamentary elections in the UK, the USA, India, and Canada, but hardly anywhere outside the former British Empire, and known in the UK as 'first-past-the-post'. To elect a single person where there are two or more candidates, you simply choose the

The Mathematical World of Charles L. Dodgson (Lewis Carroll). Robin Wilson and Amirouche Moktefi.
Oxford University Press (2019). © Oxford University Press 2019.
DOI: 10.1093/oso/9780198817000.001.0001

one with the most votes, even when they add up to fewer than half of the votes cast. As Dodgson observed, it is easy to demonstrate the 'extraordinary injustice' of this method: a candidate may win a plurality of votes – that is, more than any of the others – while being ranked last by more than half of the voters. Here is Dodgson's example (Table 1):[4]

Let us suppose that there are eleven electors, and four candidates, *a, b, c, d*; and that each elector has arranged in a column the names of the candidates, in the order of his preference; and that the eleven columns stand thus:–

a	*a*	*a*	*b*	*b*	*b*	*b*	*c*	*c*	*c*	*d*
c	*c*	*c*	*a*	*a*	*a*	*a*	*a*	*a*	*a*	*a*
d	*d*	*d*	*c*	*c*	*c*	*c*	*d*	*d*	*d*	*c*
b	*b*	*b*	*d*	*d*	*d*	*d*	*b*	*b*	*b*	*b*

Here *a* is considered best by *three* of the electors, and second by all the rest. It seems clear that he ought to be elected; and yet, by the above method, *b* would be the winner – a candidate who is considered *worst* by *seven* of the electors!

Ramon Llull was the first person to see this difficulty, and also to propose solutions. He considered the case of a monastic house where the monks or nuns must elect their own leader. As we shall see, as the first Westerner to apply algebra to voting, he invented a matrix notation for tournaments in which each candidate is compared with each other in turn.[5]

The only known copy of one of Llull's papers was made by the next great figure, Nicolaus Cusanus (Nicholas of Cusa), whose scheme differed from Llull's. Cusanus proposed that Electors to the position of Holy Roman Emperor (who were then mostly German princes) should hand in anonymous ballot papers, marking 1 for the candidate they least wanted, up to 10 (if there were ten candidates) for their favourite. We now know this as the *Borda count*, from its second proposer over three centuries later.

During the Revolutionary years Borda and Condorcet were contemporaries, but enemies, in the French Academy of Science: Borda survived, while Condorcet died in the Terror of 1794. While Borda again proposed the rank-order count, Condorcet observed that, although simple, it could produce perverse results. Borda himself found out one of his scheme's perversities: when used for electing new members of the Academy of Sciences, his scheme was manipulated by voters placing last the most dangerous rival to their favourite: 'My scheme is intended only for honest men', he said plaintively.[6]

Instead of the Borda rank-order count, Condorcet proposed a scheme to look for what is now called the *Condorcet winner*: the candidate who beats each of the others in the exhaustive pairwise tournament first proposed by Llull. Unfortunately, as he went on to discover, there is no Condorcet winner when there is a cycle at the top, because each

Ramon Llull and Nicolaus Cusanus

The Marquis de Condorcet and Jean-Charles de Borda

candidate loses to at least one other. Condorcet had a tie-break procedure, but it was not understood for a further two centuries.[7]

Dodgson and voting theory

Rather unusually in intellectual history, we can be sure that none of this prior work was known to Dodgson in December 1873 when he wrote his first seminal paper, *A Discussion of the Various Methods of Procedure in Conducting Elections.*[8] Llull and Cusanus were not rediscovered in this context until the 1980s.[9] As for Borda and Condorcet, Christ Church Library in Oxford has a run of the proceedings of the *Académie Royale des Sciences* dating back to its 17th-century foundation, but the pages in the 1781 volume that contain Borda's method are uncut: nobody has ever read them in their copy.[10] But whereas the Christ Church Library evidence is still there, another piece of evidence, discovered in the 1940s by Duncan Black, has disappeared. Christ Church has no copy of Condorcet's *Essai* of 1785, but Oxford's main university library, the Bodleian Library, has, and in Black's time some of its pages, too, were not cut.[11] (They have been now.)

The occasion for Dodgson's 1873 paper was the election of two new Students of Christ Church. Dodgson reviewed methods used for such elections and proved that all those in common use were faulty. His summary dismissal of 'first-past-the-post' was presented in Table 1.

One important point that Dodgson made was that 'no election' should also be treated as if it were the name of a candidate. The processes of starting with the question 'Should we elect anyone?', and of finishing with 'Should we elect the candidate who has won the most votes?', were both unfair to any voter who preferred one set of candidates to 'no election' but who preferred 'no election' to another set of candidates. This simple insight of Dodgson's is still too rarely understood by appointing committees.

Dodgson went on to discuss a knockout tournament, which he called the 'method of elimination'. He proved that it may lead to the 'preposterous' conclusion that the result of an election can depend on the order in which pairs of candidates are voted on, and gave a four-candidate example in which the chance of different pairs being drawn first could lead to any of the four winning.

He next discussed the 'method of marks', in which the voters have a fixed number of points to distribute among the candidates as they wish. Using reasoning that would later be considered as *game theory*, Dodgson showed that this would very quickly degenerate into simple plurality voting, which (as he had shown earlier) can elect an absolute majority loser:[12]

Each elector would feel that it was *possible* for each other elector to assign the entire number of marks to his favourite candidate, giving to all the other candidates zero: and he would conclude that, in order to give his *own* favourite candidate any chance of success, he must do the same for him.

This was an unknowing echo of Borda's 'My scheme is intended only for honest men', which is ironic because Dodgson went on to recommend precisely the Borda count, with 'no election' treated as if it were a candidate.

The college did indeed use Dodgson's modified Borda procedure, but a rather disturbing thing happened. The voting records, discovered by Duncan Black, show that the Borda winner (Becker) received 48 points, but lost a direct pairwise vote against the

Borda runner-up (Baynes), who received 47 points.[13] The college elected Baynes, and Dodgson found that his Borda count would have elected a Condorcet non-winner.

As well as electing candidates to fellowships, the college had to decide on a new belfry for the south-east corner of its main quadrangle, in a sensitive site between the cathedral and the Great Hall. Henry Liddell, Dean of Christ Church, was an architectural enthusiast who liked designing belfries. Opinion was sharply divided, in more than one direction (materials?/look?/size?), and Dodgson was openly scornful of Liddell's ideas. Immediately before a meeting of the Governing Body to discuss belfries, which (as Black discovered) was to last five hours, Dodgson rushed out his second pamphlet, *Suggestions As to the Best Method of Taking Votes, Where More Than Two Issues Are to Be Voted On*.[14] He rescinded his support for the Borda count, and instead proposed an exhaustive pairwise vote – in other words, a Condorcet comparison – reserving the Borda count for the case where the Condorcet comparison failed to produce a clear answer.

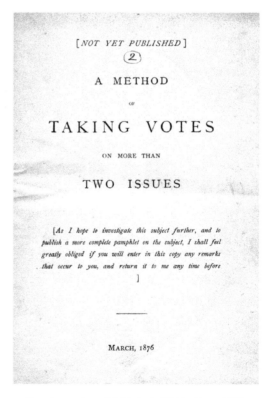

SUGGESTIONS

AS TO THE BEST METHOD OF

TAKING VOTES,

WHERE MORE THAN

TWO ISSUES ARE TO BE VOTED ON.

OXFORD:
BY E. PICKARD HALL AND J. H. STACY,
Printers to the University.
1874.

[*NOT YET PUBLISHED*]
②

A METHOD
OF

TAKING VOTES

ON MORE THAN

TWO ISSUES

[*As I hope to investigate this subject further, and to publish a more complete pamphlet on the subject, I shall feel greatly obliged if you will enter in this copy any remarks that occur to you, and return it to me any time before*]

MARCH, 1876

Title page of *Suggestions As to the Best Method of Taking Votes*, 1874

Preprint version of *A Method of Taking Votes*, 1876 [Courtesy of Edward Wakeling]

Dodgson went on thinking about these matters. His third pamphlet, *A Method of Taking Votes on More Than Two Issues*, appeared in March 1876. It was not produced in the shadow of an urgent vote, and is all the better for that. It starts with a set of Dodgsonian rules – very precise and intricate, and mystifying to anyone not as steeped in the subject as the author. An *absolute majority winner* is a candidate who (or an option which) wins the first preference of more than half of the voters. As before, Dodgson first proposed that if one outcome is an absolute majority winner then it should be selected immediately.

The next step is to check whether a Condorcet winner exists, by conducting a full pairwise tournament. If there is a winner, then the process stops. If not, then there must be a 'top cycle', which might embrace all the options. Any options that are not in the top cycle may be discarded. To break the cycle, Dodgson had a new proposal:[15]

When the issues to be further debated consist of, or have been reduced to, a single cycle, the Chairman shall inform the meeting how many alterations of votes each issue requires to give it a majority over every other separately.

Armed with this information the voters vote again, to see whether that breaks the cycle. If not, then there is no election.

To some people the existence of cycles is obvious; to others it comes as a great surprise. Here is an example of Dodgson's (Table 2):[16]

Let us suppose that there are 11 electors, and 4 candidates, *a, b, c, d*; and that each elector has arranged in a column the names of the candidates in the order of his preference; and that the 11 columns stand thus:—

a	a	a	a	b	b	b	c	c	c	d
d	d	b	b	c	c	d	b	b	b	c
c	c	d	d	a	a	c	d	d	d	b
b	b	c	c	d	d	a	a	a	a	a

Here the majorities are cyclical, in the order *a d c b a*, each beating the one next following.

To take this more slowly than Dodgson did, we observe that *a* beats *d*, *d* beats *c*, and *c* beats *b* (each by 6 votes to 5), but *b* beats *a* (by 7 votes to 4). So we can write $a > d > c > b > a$: here the symbol '>' is to be read as 'beats'. It may be easier to read this off from a vote matrix, as we show later.

'Majority rule' has a clear meaning when only two candidates or issues are in contention. If one option wins more than half of the votes cast, then it is unambiguously the majority's choice. Simple majority rule satisfies some classically desirable properties of fairness; furthermore, it was proved by Kenneth O. May in 1952 that it is the *only* voting procedure that does so.[17]

But as soon as there are more than two candidates, simple majority voting may perform very badly. For instance, a candidate may be the plurality winner (getting the largest number of votes of any candidate, but fewer than half of the votes cast), while being the absolute majority loser (being ranked last by more than half of the voters). Dodgson's example in Table 1 was designed to show this.

An obvious way to avoid this difficulty is to conduct exhaustive pairwise voting, where each candidate is compared directly with each of the others. If the voters are asked to rank the candidates, then this may be done at the second stage in the procedure, the 'aggregation stage' – there is no need for the voters actually to vote on each pair. Exhaustive pairwise voting was first proposed by Ramon Llull in 1283. If it yields a clear winner – a candidate who has won a majority against every other candidate, taken one at a time – then that candidate (the Condorcet winner) has an obvious claim to be considered the best.

But, as Condorcet had found out ninety years earlier, Dodgson discovered that the Condorcet winner may not exist (see Table 2). When a Condorcet top cycle exists, 'majority rule' seems to have no meaning: whatever society ends up with, a majority of voters would have preferred somebody (or something) else. This has deep implications for democratic theory.

The Cusanus (or Borda) rank-order count obviates this difficulty. It was described in Borda's paper, with Condorcet's commentary, neither of which Dodgson had read. The Borda count, of which the Cusanus count is a special case, works as follows. Each voter ranks the candidates from best to worst. Each last place is scored as a and each interval between places is scored as b ($b > 0$). Where there are n candidates, this is most easily done by giving n votes to one's favourite, $n - 1$ to the next, and so on down to 1 vote for one's least-liked, as proposed by Nicolaus Cusanus (so $a = 1$ and $b = 1$), or (better) from $n - 1$ at the top to 0 at the bottom, as proposed by Borda (so $a = 0$ and $b = 1$); there are rules for equal places that need not concern us. These scores are then simply added up and the candidate with the highest aggregate score is elected.

This is all beautifully simple and (unlike the Condorcet procedure) always gives a clear result. But it may fail to select the Condorcet winner, even when one exists. It may also lead to peculiar outcomes when the set of candidates is expanded or contracted, and it can patently be manipulated by voters placing the most dangerous rival to their favourite at the bottom of their lists.

In his exposition Dodgson proposed a matrix in which each candidate is scored against each other. This revived a similar notation that, unknown to him, had been proposed by Llull in 1283: one of Llull's half-matrices is shown below. Such a half-matrix is all that is needed for entering data, if two numbers are inserted into each cell.

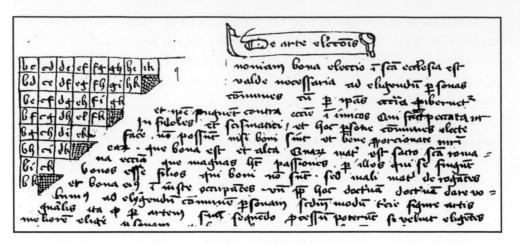

One of Llull's half-matrices

But for computation it is easier to construct a square matrix, in which the votes for *a* against *b* are entered in cell *ab* and the votes for *b* against *a* are entered in cell *ba*; the matrix corresponding to Table 2 is:

$$\begin{array}{c c c c c} & a & b & c & d \\ a & \begin{pmatrix} 0 & 4 & 4 & 6 \\ b & 7 & 0 & 5 & 8 \\ c & 7 & 6 & 0 & 5 \\ d & 5 & 3 & 6 & 0 \end{pmatrix} \end{array}$$

It is then easy to carry out arithmetic checks to confirm that the numbers have been entered correctly. In particular, when there are *m* candidates and *n* voters the cells in the matrix should add up to $(m-1)!\, n$.

If the votes for each candidate are scored against each other in such a square matrix, then the Borda and Condorcet winners may both be read directly from the matrix. If Borda scores are assigned with $a = 0$ and $b = 1$, with 0 for the least-liked of *n* candidates and $n-1$ for the voter's favourite, then they also record the number of victories that each candidate has over the others in an exhaustive tournament; Borda noticed this equivalence himself. The Condorcet winner is computed by comparing each cell with its mirror image across the main diagonal; this main diagonal is empty because an option's score against itself has no meaning. So the Borda scores for each candidate are horizontal sums of their pairwise votes. In these small elections the Condorcet comparison is done by eye, with each cell compared with its mirror image. In larger elections it is easily programmable.

The Dodgson matrix is a handy device that may be used in any context where we want to decide the Condorcet and Borda winners, provided that we have a rule that tells us what to do when they are different. Dodgson proposed two different tie-breaks, one in his 1874 pamphlet and a different one in 1876.

Lawn tennis tournaments

Exhaustive pairwise voting, as discussed by Llull and Dodgson, is essentially a tournament in which every player plays each other once. It is therefore not surprising that Dodgson next turned his attention to the new sport of lawn tennis, after watching a game in the Oxford University Parks in 1880.

For a large number of players, the number of pairs may be unmanageably large. If there are n candidates, then the number of pairs is $n(n-1)/2$, and in the 1880s a lawn tennis competition (such as at Wimbledon, where the All-England competition began in 1877) typically had 32 players, requiring an unmanageable 496 matches.[18] Before Dodgson, the authorities running the evolving sport had favoured a knockout tournament, in which each player plays one other in the first round, after which half of the players are eliminated. In each subsequent round, the number of players still in competition halves again as losers drop out, until the final is held between the last two survivors.

But a knockout system has a fatal flaw which Dodgson had already identified in 1873 in the context of voting among candidates. In his pamphlet *Lawn Tennis Tournaments*, published in 1883, he opened with the following words:[19]

At a Lawn Tennis tournament, where I chanced, some while ago, to be a spectator, the present method of assigning prizes was brought to my notice by the lamentations of one of the Players, who had been beaten (and had thus lost all chance of a prize) early in the contest, and who had the mortification of seeing the 2nd prize carried off by a Player whom he knew to be quite inferior to himself.

For, suppose that the 32 players in the tournament had been paired initially in descending order of ability (see overleaf). Then, in the first round, all even-numbered players are eliminated. In the second round, player 1 eliminates player 3, player 5 eliminates player 7,..., and player 29 eliminates player 31. So the process proceeds until the final match takes place between the best player and the 17th-best. Because numbers 2 to 8 will all have been eliminated, the third prize will be awarded to the 9th-best, and the fourth prize to the 25th-best.

In consequence, Dodgson asserted that 'the present method of assigning prizes is, except in the case of the first prize, entirely unmeaning', and proposed to lay before the reader 'A system of rules for conducting Tournaments, which, while requiring even less time than the present system, shall secure equitable results'.

Nowadays, this problem is dealt with by 'seeding' players according to their performance in earlier tournaments, and arranging that players with a similar seeding play one another: this mitigates the problem, but does not eliminate it. Dodgson's solution is more interesting: his core idea was this:[20]

A list is kept, and against each name is entered, at the end of each contest, the name of any one who has been proved superior to him — whether by actually beating him, or by beating some one who has done so (thus, if A beats B, and B beats C, A and B are both "superiors" of C). So soon as any name has 3 "superiors" entered against it, it is struck out of the list.

Over a century later, social choice-theorists rediscovered the concept, which is now called 'covering' and which helps break some Condorcet

A tennis tournament with 32 players

Let the players be arranged alphabetically, and call them A, B, C, etc., and let their relative skill be represented by the following numbers :—

A	B	C	D	E	F	G	H	J	K	L	M	N	P	Q	R
6	10	13	5	15	9	7	12	1	14	3	2	8	16	11	4

These numbers will enable the reader to decide which will be the victor in any contest : but of course they are not supposed to be known to the Tournament Committee, who have nothing to guide them but the results of actual contests. In the following tables, " I. (e) " means " first day, evening," and so on : also a player, who is *virtually* proved superior to another, is entered thus " (A)." The victor in each contest is marked *.

TABLE I. (PAIRS.)

I. (r)	II. (m)	(e)	III. (m)	(e)	IV. (m)	(e)

Part of a letter by Dodgson, published on 1 August 1883, in the *St James's Gazette*

cycles.[21] Where there is no Condorcet winner, there may still be an 'uncovered set' in which the plausible winners may be found.

Proportional representation

Until 1883 all of Dodgson's work on choice procedures addressed the case of choosing a single winner (person or outcome), to be selected by a tournament method. The problem of proportional representation is different and emerges from the *microcosm theory*. This theory for representative bodies holds that the body should represent those who elect it, in such a way that each relevant characteristic of the electors (such as age, gender, ethnicity, or political opinion) is represented in the assembly in the same proportion as it appears in the electorate. However, this concept of proportionality has no meaning in (for instance) a presidential election: there is only one president, who cannot be white, black, gay, straight, female, male, left, and right in the same proportion as the electorate. So the study of proportional representation (PR) is a distinct branch of the mathematics of voting.[22]

Here, unlike in the single-winner case, Dodgson knew some of the prior and contemporary work. From his perspective, first-past-the-post was bad enough, but the actual UK parliamentary electoral system was worse. Many UK constituencies, including Oxford University, returned more than one member, and this only exacerbated the problem of first-past-the-post that Dodgson had identified in 1873. If an absolute majority loser could win one seat, then two absolute majority losers could win each of two seats.

Various enthusiasts, publicized by John Stuart Mill, had been proposing PR schemes in Britain for most of the 19th century, and even implementing them in some colonial assemblies.[23] In four letters to the *St James's Gazette*,[24] Dodgson showed that some of their ideas were mathematically unsound. In particular, the system boosted by Mill, and now known world-wide as the 'single transferable vote' (STV), may elect the wrong set of candidates.

STV depends on successive eliminations and transfers of surpluses, but the latter are a minor problem. The eliminations are a major problem, however, for the reason that Dodgson had already identified – that a top cycle member or Borda winner may be eliminated early on, like the unlucky knockout victim of a tennis tournament. Dodgson's own proposals were hard to follow but, as with his work on social choice, they are now recognized as unique.

In Dodgson's day the widening of the franchise in the United Kingdom was a heated issue. Conservatives knew that it was difficult to resist the principle, yet feared that they would lose out substantially. The franchise was reformed in 1832 and 1867–68 when qualifications to vote were slightly lowered, and a few 'rotten boroughs' with very small electorates were abolished and their seats redistributed to growing industrial cities. The 1867 Reform Act included a provision that some of the large English cities should each form a three-member district, with voters having only two votes each: this was known as the 'limited vote', and it was added to the bill by Lord Cairns's hostile amendment in the House of Lords which the Commons Tory leader, Benjamin Disraeli, unexpectedly accepted.[25] The impact of the limited vote on Dodgson's thinking is discussed below.

Except in Ireland, the British General Election of 1880 saw the closest approach since 1841 to a straight two-party contest, between Liberals and Conservatives. It illustrated both the bias and the exaggerative effect of the first-past-the-post electoral system. The Liberals won roughly the same proportion of seats as votes, but the Conservatives won many fewer seats than their vote share, and the Irish Party many more.

So the Liberals formed a government, and in 1884 they proposed a further widening of the franchise. The Marquess of Salisbury, leader of the Conservatives, noted the bias and exaggeration in the 1880 result. He envisaged that, unless the extension of the franchise were accompanied by a redistribution of seats, the Conservatives could be decimated in Parliament, even if the franchise extension reduced their share of the vote only slightly (or not at all). In the most sophisticated piece of election analysis yet written by a UK parliamentary leader, he showed that if the electorate of a 17-seat legislature with single-member districts were split between imaginary parties (which he named 'Catholics' and 'Liberals') in the proportion 8 to 9, then the 'Liberals' would win all 17 seats in two cases:[26] either where the population was exactly evenly mixed, *or* where it was completely segregated (say, into a 'Liberal' city surrounded by 'Catholic' countryside), but where constituencies were drawn radially from the city centre in such a way that each constituency contained the same ratio of 'Catholics' to 'Liberals' as the population.

Ireland was even more threatening to Salisbury's interests. The whole of Ireland was incorporated into the United Kingdom in 1801, but the union was never legitimate in the eyes of Irish Catholics, who comprised some seven-eighths of the population.[27] Since 1874, seats in Catholic Ireland had been falling to supporters of Home Rule, who used every procedural means open to them to disrupt Parliament. The franchise reform of 1884 proposed to give the vote in Ireland, as in the rest of the country, to rural male householders. But would this not mean a great boost to Charles Stuart Parnell, the Home Rule leader, with consequent threats to public order and the unity of the UK? As Sir John Lubbock, one of the leading advocates of STV, wrote:[28]

At the general election of 1880, 86 seats were contested [in Ireland]. Of these the Home-rulers secured 52, the Liberal and Conservatives together only 34. Yet the Home-rule electors were only 48,000, while the Liberals and Conservatives together were no less than 105,000...we are

Lord Salisbury, Lord Cairns, and Sir John Lubbock

told…that under the new Redistribution Act the Home-Rulers will secure 90 seats out of 100, leaving only a dozen to the Liberals and Conservatives together…The result of this system, then, will be that Ireland will be entirely misrepresented, and that we shall have gratuitously created serious and unnecessary difficulties for ourselves. To adopt, indeed, a system of representation by which we shall exclude from the representation of Ireland one-third of the electors, and give almost the whole power to two-thirds, would, under any circumstances, be unjust; but to do so when the one-third comprise those who are moderate and loyal, while the two-thirds are led by men not only opposed to the Union, but in many cases animated by a bitter and extraordinary hatred of this country, seems to be an act of political madness.

He then went on to draw attention to the US Presidential election of 1860, in which Abraham Lincoln won an absolute majority of the Electoral College on less than 40 per cent of the vote.

Lubbock's prediction was spot on. The Home Rulers won 85 seats in the 1885 General Election, and continued to do so at every election until 1910. Anglo-Irish war was about to break out in 1914, but was delayed for the First World War to take place. The Irish wars lasted from 1919 to 1923. Electoral systems have important consequences.

This was the context for Dodgson's *Principles of Parliamentary Representation* of 1884. Dodgson was a political Conservative as well as a temperamental conservative. He met Salisbury and his family in 1870, uncharacteristically using his fame as the author of *Alice* to obtain an introduction to Salisbury's wife and daughters. Despite the gulfs of class and temperament, he was welcomed by the Salisbury family and spent the New Year at their great house, Hatfield, several times during the 1870s and 1880s.

Dodgson seems to have thought about PR for the first time in 1882, in connection with college politics, but it was the reform crisis of 1884 that brought him into print. His letters to the *St James's Gazette* show his continually evolving ideas. In June he hit on the most distinctive feature of his scheme, 'the giving to each candidate the power of transferring to any other candidate the votes given for him', and in the following month he sent it to Salisbury, saying:[29]

How I wish the enclosed could have appeared as *your* scheme…That *some* such scheme is needed, and much more needed than *any* scheme for mere redistribution of electoral districts, I feel sure.

Salisbury replied immediately, acknowledging the need for electoral reform but stressing the difficulty of getting a hearing for 'anything…absolutely new…however Conservative'. Dodgson responded the next day. After congratulating Salisbury for insisting that the Conservatives (who controlled the House of Lords) would not accept franchise reform unless it were linked with redistribution, Dodgson continued:

please don't call my scheme for Proportionate Representation a 'Conservative' one…all I aim at is to secure that, *whatever* be the proportions of opinion among the Electors, the same shall exist among the Members.

The House of Lords did indeed return the franchise bill to the Commons with an added clause insisting that it must be 'accompanied by provisions for so apportioning the right to return members of parliament as to insure a true and fair representation of the people'. This amendment was moved by Lord Cairns, the same peer who had inserted the limited vote provision into the 1867 Reform Act. No doubt the motives of the majority Conservative peers mingled self-interest with a desire for fair representation. In Britain, fair representation was expected to mean protecting the Conservatives from being wiped out in terms of seats in a General Election where they narrowly came second in terms of votes. Salisbury may have wished to preserve the limited vote, but in the end he went down a different road.

Like Lubbock, Dodgson saw that Salisbury had failed to accept the implications of his own argument. No redistribution based on single-member districts with the plurality voting rule could be guaranteed to save the Conservatives in Britain, nor either British party in Ireland. As Salisbury had pointed out in the *National Review*, and as Dodgson repeated in his *Principles of Parliamentary Representation*, single-member districts when combined with an even distribution of supporters of two parties around the country could lead to the larger of the two wiping out the smaller in terms of seats. This happened in Ireland in every election from 1885 to 1910, and in Scotland in 2015.

Salisbury could not shift his perspective from majoritarian to proportional in order to see the true implications of his own argument. The concepts of political and physical proportionality interact with one another. Conservative supporters were fairly evenly spread around Great Britain; Irish Nationalists were to be found only in Ireland (and in Liverpool, where they held a seat from 1885 to 1929). So he turned to a scheme that achieved physical proportionality in Britain, although not in Ireland, and masterminded the scheme of single-member districts 'according to the occupation of the people' that was embodied in the Redistribution of Seats Act of 1885. This scheme, the outcome of his negotiations with the Liberals' Sir Charles Dilke in the autumn of 1884, has frequently been hailed as Salisbury's stroke of genius. It constructed suburban seats where the new concentration of Conservative voters was to be found in what contemporaries called the 'villa vote',[30] but it is unclear whether it would have saved the Conservatives if the debacle that Salisbury feared had come about. For in 1886 the Liberals were torn apart over Irish Home Rule, leading to a twenty-year Conservative hegemony not foreseen by Salisbury or anybody else in 1884. Thus, in the end, no Conservative had to take Dodgson's arguments seriously out of self-interest.

It is unfortunate that certain points that Dodgson quickly passed over in his *Principles of Parliamentary Representation* were exactly the ones that mainstream politicians could not accept, even when it was in their own interest. Dodgson took for granted that guaranteeing the survival of minorities in parliament requires multi-member districts and minority representation (which the politicians should have accepted, but did not), and that the number of electors per Member of Parliament should be equal (which almost no parliamentarian in the 1884 debates did). These assumptions were probably enough on their own to blind contemporaries to the more striking features of the *Principles*.

The *Principles* is the earliest known work to discuss both the assignment of seats to each of a number of multi-member districts (the *apportionment problem*) and the assignment of seats within each district to the parties (the *PR problem*). Not until 1982 was the formal congruence of the two problems fully understood.

Dodgson's own preference was for electoral districts with two to five members, in which electors were given just one vote. For such multi-member constituencies Dodgson calculated the percentages of the votes required for a political party to return a specified number of members, and presented his results in the following table:

		number of seats to fill					
		1	2	3	4	5	6
number of members allocated to district	1	**51**					
	2	34	**67**				
	3	26	51	76			
	4	21	**41**	61	**81**		
	5	17	34	51	67	84	
	6	15	29	43	58	72	86

For example, in a single-member district a party requires over half of the votes in order to fill the seat – that is, at least 51 per cent (line 1, column 1), if we stick to whole numbers. In order to return two members, a party requires at least 67 per cent of the vote (line 2, column 2) in a two-member district, and at least 41 per cent (line 4, column 2) in a four-member one. In general, as Dodgson discovered, a political party wishing to return k members in an n-member district requires more than $k / (n + 1)$ of the votes; for example, in order to fill two seats ($k = 2$) in a four-member district ($n = 4$), a party needs more than two-fifths of the votes – that is, at least 41 per cent. The last number in each line gives the percentage of the electorate represented by its members; for example, in a four-member district, 81 per cent of the electorate is so represented.

Thus, for the assignment of seats to parties, Dodgson essentially recommended what came to be known as the standard 'Droop quota' $V / (S + 1)$, rounded up to the next integer, where V is the total number of votes cast and S is the number of seats to be filled. In 1881 the mathematician Henry R. Droop had pointed out that a higher quota proposed by Thomas Hare and endorsed by Mill could lead to too few candidates being elected. Dodgson showed that the rules of the Proportional Representation Society (now the Electoral Reform Society) to run a single transferable vote could lead to the defeat of a candidate who had obtained a Droop quota. This cannot happen at the first stages of an STV election, but Dodgson showed that it can happen at subsequent stages. He briskly concluded:[31]

I think I have sufficiently proved the fallacy of its method for disposing of surplus votes…Clearly *somebody* must have authority to dispose of them: it cannot be the Elector (as we have proved); it will never do to refer it to a Committee. There remains *the Candidate himself, for whom the votes have been given.*

Henry R. Droop
[The Master and Fellows of Trinity College, Cambridge]

Dodgson's surprising formula does not get to the root problem of STV and all other elimination systems – namely, that they use information about voters' preferences after their first in an arbitrary way. Preference orderings are not treated equally: the $(n + 1)$th preference of a voter whose nth preference has been eliminated is counted, whereas the $(n + 1)$th preference of a voter whose nth preference has been elected with a surplus is counted with reduced weight, and the $(n + 1)$th preference of a voter whose nth preference is elected with nothing to spare is not counted.

Dodgson presented his alternative in a compressed and elliptical argument. He identified what game theorists now call the *Nash equilibrium strategy* for two parties, after the Nobel Prizewinner John Nash. A *Nash equilibrium*, not so defined until 1951, is a point in a game from which, once reached, it benefits neither player to depart. Dodgson considered the class of methods in which voters may each cast v unranked ballots in an m-member district, where $v < m$. The limited vote introduced by Lord Cairns operated in Manchester, Birmingham, and Liverpool with $v = 2$ and $m = 3$, and in the City of London with $v = 3$ and $m = 4$. In Birmingham the Liberals had manipulated the limited vote by dividing the city into three zones and asking their supporters in each zone to vote for a different pair from the Liberal candidates: they thereby won all three seats in each of the general elections in the period. In 1874 the Conservatives did not run at all. In 1868 the Liberals controlled 73% of the votes cast, and in 1880 they controlled 67%.

Were the Birmingham parties rational strategists? This may be answered directly from Dodgson's pamphlet, once his reasoning is understood. Given perfect information and common knowledge about party strengths, Dodgson knew, but failed to make explicit, that the optimum strategy for each party is to put up exactly as many candidates as it can

John Nash
[Wikimedia Commons]

fill seats, and to instruct its supporters to divide their ballots among its candidates as evenly as possible.

Dodgson's aim was to find the voting procedure that leaves the fewest voters 'unrepresented' by votes that do not contribute to the outcome. He concluded that, within the class of limited voting procedures, the fairest method – in the sense that it can be predicted to leave the fewest voters unrepresented – is that where $v = 1$ for any m. The fairness of the system increases with m, so the fairest of this class of systems is that which divides the country into multi-member constituencies in which each voter has only one vote. This was the system used in Japan for national elections from the end of the Second World War until 1993. It was generally labelled the *single non-transferable vote* (SNTV), and has also been used in Taiwan and Korea. In these countries, m has typically had a value of around 5.

From Dodgson's calculations we find that, for $v = 2$ and $m = 3$, a party with at least 61% of the vote can guarantee to win all three seats, so in Birmingham the Liberal strategy was optimal. The Conservative strategy was optimal in 1874, and was sub-optimal in 1868 and 1880, at any rate from the perspective of Dodgson's game. If it was common knowledge that the Liberals controlled over 60% of the probable voters, then there was no point in the Conservatives' running candidates, unless for the purpose of forcing an election and putting the Liberals to some expense.

Dodgson's result on SNTV has been independently rediscovered by Gary Cox and Emerson Niou.[32] It can be obtained by postulating either that the parties are rational (in which case they do what the Birmingham Liberals did) or that the voters are rational (in which case, given perfect and free information, they would clump their ballots in the same way as the Birmingham Liberals directed them to). The outcome of a procedure in which the parties seek to minimize waste of 'their' votes is the same as that of a procedure in which the voters seek to avoid wasting their votes.

Neglect and rediscovery

Many writers have constructed elaborate psychological theories to account for Dodgson's lack of recognition during his lifetime, in anything except *Alice* and *Snark*. However, nothing elaborate is required to explain Dodgson's lack of immediate influence in college, university, or world affairs. He could be pedantic, quarrelsome, and (when not quarrelsome) obscure.

By the time of the belfry controversy Dodgson had fallen out acrimoniously with Dean Liddell, publishing sarcastic pamphlets about him which, to a modern reader, have none of the lightness of the *Alice* books.[33] His correspondence with Lord Salisbury (which we reviewed above) shows that Dodgson was not always quarrelsome and that he could engage with a senior politician – indeed, one of the most influential politicians in the UK. It is a pity that he could not do so in the smaller world of Christ Church.

Even there, it has been argued that Dodgson had agenda-setting, or rule-setting, power. In the light of his behaviour towards Liddell it seems surprising that his colleagues ever listened to his advice, but we know that they did – for instance, by using the Borda count in the election of Baynes in 1873. Although that was not an unmitigated triumph (because Baynes was not the Borda winner), it spurred Dodgson on to his most fundamental work on voting.

Alice Adventures in Wonderland and *Through the Looking-Glass* are based on games, the latter being explicitly a game of chess. Dodgson's writings about voting were also about games, as Duncan Black was the first to see. Indeed, Dodgson's life was devoted to being serious about games and game-like about serious things. Occasionally he actually referred to a voting 'game', as when proving that there was what would now be called a *defective equilibrium* when rational voters used his 'method of marks'. But his more remarkable achievement was to have written about voting in game-theoretic terms, before game theory had been invented.

Alice is deep; *The Hunting of the Snark* is deep and dark; Dodgson's writings on voting and proportional representation were also deep, and were not understood for almost a century.

At a pace of six miles in the hour: from *Excelsior* (Knot I of *A Tangled Tale*)

CHAPTER 6

Recreational mathematics

EDWARD WAKELING

> Yet what are all such gaieties to me
> Whose thoughts are full of indices and surds?
>
> $$x^2 + 7x + 53$$
> $$= \,^{11}/_3$$
>
> C. L. Dodgson, *Four Riddles*, 1857

To suggest that Dodgson's recreational games and puzzles trivialized his mathematical achievements would be mistaken. To suggest that Dodgson's aim in inventing mathematical puzzles was simply to entertain children would also be mistaken. Dodgson had a lifelong interest in recreational mathematics, enjoying the creations of others and inventing many of his own to test and amuse his friends and colleagues.

Introduction

Any activity in which mathematics is used to amuse and entertain may be described as recreational mathematics, in some cases as a relaxation or distraction that can be agreeably indulged in and used to refresh the mind. No specific standard is imposed on the mathematical problems and puzzles, but many are aimed at the non-mathematician, thereby making them accessible to all.

The Mathematical World of Charles L. Dodgson (Lewis Carroll). Robin Wilson and Amirouche Moktefi.
Oxford University Press (2019). © Oxford University Press 2019.
DOI: 10.1093/oso/9780198817000.001.0001

Many of Dodgson's puzzles and games are concerned with language and logic, but these are not within the remit of this chapter. Our concern here is mainly with numerical problems and those linked to other branches of mathematics. Dodgson had a particular fascination with puzzles concerning probability, but arithmetical and geometrical puzzles were also of interest to him.

This chapter also includes some of the useful mathematical rules invented by Dodgson to work out key results, such as the day of the week for any date and methods for testing divisibility in numbers. Dodgson saw these rules as examples of recreational mathematics and intended to include them in a publication on this topic. These days examples of recreational mathematics abound, often printed daily in newspapers, with Sudoku as a continuing favourite.

It is impossible to include in this chapter all the puzzles and problems that Dodgson invented, so a selection has been made. Other sources of examples will be found in the *Further reading* section at the end of this book.

Using puzzles and games to entertain

Puzzles and games continued throughout Dodgson's life, mainly as a form of entertainment to amuse himself, his colleagues, and his friends. One of his earliest games, created in 1855 when he was 23 years old and mentioned in his *Diaries*, was 'Ways and Means'.[1] This was a card game based on auctions and requiring buying and selling using counters, and is principally a word game involving strategy. Dodgson played it with his siblings and visitors: it still exists, but it has never been published.

As we saw in Chapter 1, soon after Dodgson took on the role of Lecturer in Mathematics at Christ Church, he offered to teach 'sums' at St. Aldate's elementary school, just opposite Tom Gate in Oxford. All began well with the first few lessons undertaken in January 1856, but soon Dodgson was describing the lessons as 'noisy' and 'unmanageable.' He resorted to a few mathematical tricks to gain the attention of the pupils.[2] On 5 February he tried the puzzle

of writing the answer to an addition sum, when only one of the five rows has been written: this, and the trick of counting alternately up to 100, neither putting on more than 10 to the number last named, astonished them not a little.

Then on 8 February he showed them 'the "9" trick of striking out a figure, after subtracting a number from its reverse', a well-known puzzle that can be traced back to the Middle Ages. These three puzzles, and their explanations, are described in more detail here; none of them was invented by Dodgson.

Such mathematical 'tricks' did not succeed in gaining the pupils' attention in the long term. Dodgson gave up the experiment by the end of February, and never again offered to teach mathematics to young school pupils. However, for the rest of his life the use of mathematical puzzles became a feature of his entertainment when engaging with small

Three puzzles for St Aldate's

Addition sum

In this puzzle you give me any four-digit number, say 2879; I then write down another number (22877) on a sheet of paper and hide it away. We now take it in turns to propose four more numbers: for example, you choose 4685, I choose 5314, you choose 7062 and I choose 2937. If you now add these five numbers together, you will find that their total is indeed 22877, the number I predicted. How did I arrange this?

Answer: Note that $5314 = 9999 - 4685$ and $2937 = 9999 - 7062$; in general, I subtract from 9999 whatever numbers you choose. The final total is then the original number (2879) plus (9999 + 9999), which is 22877, as predicted.

Counting alternately up to 100

In this trick we start with the number 1 and then take it in turns to add a new number, never exceeding 10; the person that reaches 100 is the winner. For example, if to the 1 you first add 4 (giving 5), I'll then add 7 (giving 12); you might then add 8 (giving 20), and I'll then add 3 (giving 23). Continuing in this way we might get the sequence 29, 34, 43, 45, 48, 56, 57, 67, 75, 78, 86, 89, 92, 100, and I win! How can I ensure that I am *always* the one to reach 100?

Answer: Note that $7 = 11 - 4$ and $3 = 11 - 8$; in general, I subtract from 11 whatever number you choose and add the result to the total. Thus the numbers I announce increase by 11 each time: 1, 12, 23, 34, 45, 56, 67, 78, 89, and 100.

[Dodgson later developed this idea into a game called *Arithmetical Croquet*.]

The '9' trick

Here I ask you to choose any number, reverse it, and then subtract the smaller number from the larger. From the answer you should then select any digit other than 0, remove it, and tell me the sum of the remaining digits: I will then tell you which number you removed. For example, if you choose 6173 and subtract its reverse, 3716, you get 2457; if you then choose to remove 5 and give me the sum of the remaining numbers $(2 + 4 + 7 = 13)$, I can then tell you immediately that it was 5 that you removed. How do I know?

Answer: If you take any number and subtract its reverse, the result can always be divided exactly by 9, and so can the sum of its digits. When you remove one digit and tell me the sum of the remaining digits, I then calculate how much I must add to increase this sum to a multiple of 9; for example, if you give me the sum 13, I must add 5 to increase the total to 18 (which is 2×9), so the number you removed was 5.

groups of young and old. Dodgson was frequently invited to visit the families of his friends and acquaintances, not just his colleagues at the University, and it became a feature of these visits that puzzles, games, and story-telling were on the agenda, continuing an activity that he had established with his own siblings.

Problems involving time

In his youth Dodgson had a fascination with the concept of time, a topic that was becoming more significant with the development of the railways and the need for accurate timetables. From the age of 11 his home at Croft-on-Tees was very near the first public railway between Stockton and Darlington, probably with the trains being heard from the Rectory where the family lived. Dodgson invented a railway game for his ten siblings to play in the Rectory garden.

Accurate timekeeping was becoming essential for trains to provide a reliable service across the country. Until then, many places worked on local time – for example, Oxford time was five minutes behind London time, being sufficiently west of the metropolis. This notion is still preserved today at Christ Church, with Cathedral evensong at 6.05 p.m. and curfew being struck by 101 rings of Tom Bell at 9.05 p.m. each evening.

Where does the day begin?

Dodgson knew that time could be stopped or fixed if you travelled west with the sun: in fact, if you travel fast enough you can make time travel backwards! Such an idea exercised his mind, particularly with global communication in its infancy and with cables being laid across land and oceans. In the days before the International Date Line Dodgson, a young man aged between 18 and 21 years, edited a family magazine entitled *The Rectory Umbrella*,[3] and in this he contributed 'Difficulties No. 1, 'Where does the day begin?'. The text begins:

Half of the world, or nearly so, is always in the light of the sun: as the world turns round, this hemisphere of light shifts round too, and passes over each part of it in succession.

Supposing on Tuesday it is morning at London; in another hour it would be Tuesday morning at the west of England; if the whole world were land we might go on tracing Tuesday Morning, Tuesday Morning all the way round, till in 24 hours we get to London again. But we *know* that at London 24 hours after Tuesday morning it is Wednesday morning. Where then, in its passage round the earth, does the day change its name? where does it lose its identity?

Dodgson continued by describing the difficulty, especially when a point is eventually reached when the name of the day changes. No doubt his siblings were amused and confused in equal measure by this contribution, but it had a serious point to make. A few years later, on 23 February 1857, he noted in his diary:[4]

Wrote a letter to the *Illustrated London News* on the subject of "Where does the day begin {?}" which I see is now being discussed in that paper.

The letter was signed 'A Mathematics Tutor, Oxford', and appeared on 18 April.

The topic resurfaced at different times in Dodgson's life. In 1885 Dodgson initiated a conversation on the topic with The Eastern Telegraph Company, Limited, of London, wanting to know how they were able to date communications around the world. He received this reply:[5]

The Secretary of the Extension Company has handed me your letter, and the only reply I can give to your question is that we cannot possibly fix the original date and time of its receipt in London. It depends on the state of the lines as regards electrical condition, pressure of traffic, ability of clerks, etc. all of which affect the speed of transmission. If you desire the theoretical time, about which we are not deeply concerned, it will be readily found by the simple process of adding or deducting the difference of times given at the end of our Tariff Book, everything after midnight belonging to the preceding or following day. We open received telegrams from Australia dated the previous day.

Dodgson was not entirely satisfied with this answer. Although this was a serious problem, the matter was not resolved until the introduction of the International Date Line which passes through oceans mainly along 180° longitude. It was initially proposed in 1884, and it has been modified several times since.

Two *Pillow-Problems*

In 1893 Dodgson compiled a series of questions that he had thought out without the use of pencil and paper over the previous twenty years, working almost all of them out mentally during the night and writing down the solution the next morning (see Chapter 1). He initially called these questions *Curiosa Mathematica, Part II, Pillow-Problems, Thought Out During Sleepless Nights*, but later changed the last phrase to *Thought Out During Wakeful Hours* to allay the unfounded fears of friends that he suffered from insomnia. The questions deal with topics in arithmetic, geometry, trigonometry, algebra, and probability, and the book first presents the *Questions*, then gives brief *Answers*, and concludes with fully written-out *Solutions*. Of these seventy-two questions, several could be described as recreational mathematics.[6]

5. (19, 31)

A bag contains one counter, known to be either white or black. A white counter is put in, the bag shaken, and a counter drawn out, which proves to be white. What is now the chance of drawing a white counter?　　[8/9/87

6. (19, 32)

Given lengths of lines drawn, from vertices of Triangle, to middle points of opposite sides, to find its sides and angles.

7. (19, 33)

Given 2 adjacent sides, and the included angle, of a Tetragon; and that the angles, at the other ends of these 2 sides, are right: find (1) remaining sides, (2) area.　　[4 or 5/89

8. (20, 34)

Some men sat in a circle, so that each had 2 neighbours; and each had a certain number of shillings. The first had 1/ more than the second, who had 1/ more than the third, and so on. The first gave 1/ to the second, who gave 2/ to the third, and so on, each giving 1/ more than he received, as long as possible. There were then 2 neighbours, one of whom had 4 times as much as the other. How many men were there? And how much had the poorest man at first?　　[3/89

9. (35)

Given two Lines meeting at a Point, and given a Point lying within the angle contained by them: draw, from the given Point, two lines, at right angles to each other, and

---

5. (2, 31)

Two-thirds.

6. (2, 32)

Calling the sides '2a', '2b', '2c', and the lines 'α', 'β', 'γ', we have

$$a^2 = \frac{-\alpha^2 + 2\beta^2 + 2\gamma^2}{9},$$

$$\triangle d = \frac{5\alpha^2 - \beta^2 - \gamma^2}{2\sqrt{2\alpha^2 - \beta^2 + 2\gamma^2}.\sqrt{2\alpha^2 + 2\beta^2 - \gamma^2}}.$$

7. (2, 33)

Let AB, AD be given sides, and B, D the right $\angle$s; and let $AB = b$, $AD = d$.

(1) $BC = \dfrac{d - b \triangle A}{\triangle A}$; $CD = \dfrac{b - d \triangle A}{\triangle A}$;

(2) area $= \dfrac{2bd - (b^2 + d^2)\triangle A}{2 \triangle A}$.

Pillow-Problems: a page of Questions and a page of Answers

Two of the Pillow-Problems, Questions 31 and 39 (presented below), involve time: one compares a watch and a clock each giving different times, and the other concerns the time taken by two people to walk along a road. Other Pillow-Problems involving probability, algebra, and geometry appear later in this chapter.

Question 31. On July 1, at 8 a.m. by my watch, it was 8*h*. 4*m*. by my clock. I took the watch to Greenwich, and, when it said 'noon', the true time was 12*h*. 5*m*. That evening, when the watch said '6 *h*.', the clock said '5*h*. 59*m*.'

On July 30, at 9 a.m. by my watch, it was 8*h*. 57*m*. by my clock. At Greenwich, when the watch said '12*h*. 10*m*.', the true time was 12*h*. 5*m*. That evening, when the watch said '7 *h*.' the clock said '6*h*. 58*m*.'

My watch is only wound up for each journey, and goes uniformly during any one day: the clock is always going, and goes uniformly.

How am I to know when it is *true* noon on July 31?

Question 39. *A* and *B* begin, at 6 a. m. on the same day, to walk along a road in the same direction, *B* having a start of 14 miles, and each walking from 6 a. m. to 6 p. m. daily. *A* walks 10 miles, at a uniform pace, the first day, 9 the second, 8 the third, and so on: *B* walks 2 miles at a uniform pace, the first day, 4 the second, 6 the third, and so on. When and where are they together?

Dodgson's answers to these two questions (supplied with full solutions) were 'When the clock says '12*h*. 2*m*. 29 $\frac{277}{288}$ *sec*.' and 'They meet at end of 2*d*. 6*h*., and at end of 4*d*.: and the distances are 23 miles and 34 miles.'

Other time problems

Another early puzzle concerning time was 'The Two Clocks':

One clock tells the time accurately twice a day, and the other tells the time twice a year. Which is the better clock to own?

This was also posed by Dodgson in *The Rectory Umbrella* for his siblings to ponder. His answer was the second clock, since the first clock is broken while the second loses a small amount each day.

Ideas of time also featured in his popular writing for children, such as in 'A Mad Tea-Party' in *Alice's Adventures in Wonderland*. Here, time was perpetually at six o'clock as a result of the Hatter's quarrelling with Time.

Problems involving probability

Probability, and betting in particular, was the focus of Dodgson's interest for a time. On 12 March 1856 he recorded in his diary:[7]

Discovered a principle (probably long known), of making a winning book on any race where *the sum of the chances* (according to market odds) *is not exactly one.* Reduce to a common denominator: put that back into odds, and make your bets in sums proportional to those numbers, *giving all* the odds if the sum of the chances *exceeds* one, and *vice versa*.

Dodgson contributed to *The Pall Mall Gazette* a paper which he entitled 'The Science of Betting' concerning this anomaly (see Chapter 7). It was published on 19 November 1866, together with a subsequent correction published on the following day. He had previously sent a letter about the same topic to *Bell's Life* in 1857.[8]

Card games

Gambling was probably frowned upon in the Dodgson household, his father being a high-church cleric, Canon of Ripon Cathedral, and Archdeacon of Richmond. Dodgson was a latecomer to card games, which were never played in the family home. He did not learn to play them until January 1858 at the age of 26, although he knew the composition of a pack of playing cards. As he recorded in his diary for 16 January 1858:[9]

Bought Hoyle's *Games*. I have taken to learning cards in the last few days, for the first time in my life.

Within nine days of learning to play cards he was busy inventing his own game for two or more players, which he called 'Court Circular'. He developed its rules over the next few months and published the game as *Rules for Court Circular* in January 1860. This was one of his first publications and appeared anonymously, although he had already written various poems and stories for magazines and journals. He continued to revise and improve the game, and published a different version in 1862. Three years later, playing cards became a major feature of *Alice's Adventures in Wonderland*, where several of the characters were based on them.

The game of Sympathy

Many manuscripts containing Dodgson's puzzles were among his surviving mathematical papers and lecture notes. At his death in 1898 these papers were acquired by Henry Tresawna Gerrans, a Fellow of Worcester College, Oxford, who had studied mathematics as an undergraduate at Christ Church from 1877 and was one of Dodgson's pupils. When Gerrans died in 1921 his widow sold the papers to Falconer Madan who had recently retired as Bodley's Librarian. In 1929 Madan sold them to Morris L. Parrish who later deposited them at Princeton University, USA, where they are now part of the Parrish Collection.

A number of these mathematical manuscripts concern a problem that had been discussed in the Senior Common Room (SCR). As the Lecturer in Mathematics at Christ Church from 1856, Dodgson was a member of the SCR, at that time a gentleman's club for the dons where many of their domestic needs were catered for, with the dons dining in Hall and retreating to the SCR afterwards for dessert and conversation. Sharing puzzles with colleagues was a feature of the entertainment, and in Dodgson's case many of these were his own invention. However, we know that puzzles being discussed in newspapers and by word-of-mouth also featured.

On 29 February 1856 Dodgson wrote in his diary:[10]

I have been trying for the last two days to solve a problem in chances, given me by Pember, which is said to have raised much discussion in the college. It is an exceedingly complicated question, and I have not yet got near a solution.

[Problem in the game of "Sympathy."

The game is this: two players lay out two separate packs in heaps of 3, (and one card over in each pack), turning each top card face upwards, so as to have 18 faces on each side. Those which correspond are paired off together, and the cards under them turned face up: (the simplest way would be, to lay all face up originally).

Required: the chance of the whole pack being paired off in this way.]

This bracketed section was written on the opposite page of Dodgson's diary where he supplemented his diary account with additional information. He also attempted to analyse the probability of success when playing the game of Sympathy:

The only cause of failure in the game is the circumstances of cards being covered on one side by cards whose fellows are covered on the other, the covering cards in the second case being fellows to the covered in the first. This may occur in many ways, and the chance of each must be taken separately.

There are several pages of calculation in which Dodgson considered the ways in which the sequence of cards blocked a solution. For example, he wrote:

One heap may "lock" another – i.e. in one heap of A's pack may be the sequence a, b, c, and in B's be b, a, c where the cards "a" and "b" can never be paired off.

Dodgson considered the chances of selecting a card from various heaps, and after voluminous calculations he did not appear to arrive at a solution: there were too many sequences to consider and the solution was too difficult to determine.

Pillow-Problems

Of the seventy-two Pillow-Problems thirteen are on probability, and ten of these involve selecting counters at random from a bag.[11] To solve such difficult problems mentally, and on the whole correctly (as Dodgson did), required extraordinary abilities. A typical problem is:

Question 5. A bag contains one counter, known to be either white or black. A white counter is put in, the bag shaken, and a counter drawn out, which proves to be white. What is now the chance of drawing a white counter?

In his solution Dodgson first observed that, as the state of the bag after the operation is the same as its state beforehand, this chance might appear to be ½. He then proceeded more carefully, calculating the probabilities at each stage, and correctly concluding:

Thus the chance, of now drawing a white counter, is ⅔.

A more complicated counters-in-bags problem is Question 50:

Question 50. There are 2 bags, *H* and *K*, each containing 2 counters: and it is known that each counter is either black or white. A white counter is added to bag *H*, the bag is shaken up, and one counter is transferred (without looking at it) to bag *K*, where the process is repeated, a counter being transferred to bag *H*. What is now the chance of drawing a white counter from bag *H*?

This is perhaps the most complex of Dodgson's thirteen probability Pillow-Problems, and Dodgson devoted over a page of calculation to give the correct solution of $^{17}\!/_{27}$; his solution was as follows.

---

50. (11, 23)

At first, the chance that bag *H* shall contain

$$2\ W \text{ counters, is } \tfrac{1}{4}.$$
$$1\ W \text{ and } 1\ B, \text{ is } \tfrac{1}{2}.$$
$$2\ B, \qquad\ \text{ is } 1.$$

∴, after adding a *W*, the chance that it shall contain

$$3\ W, \qquad\qquad \text{ is } \tfrac{1}{4}.$$
$$2\ W,\ 1\ B, \qquad \text{ is } \tfrac{1}{2}.$$
$$1\ W,\ 2\ B, \qquad \text{ is } \tfrac{1}{4}.$$

hence the chance of drawing a *W* from it is

$$\tfrac{1}{4}\times 1 + \tfrac{1}{2}\times\tfrac{2}{3} + \tfrac{1}{4}\times\tfrac{1}{3} : \text{ i. e. } \tfrac{2}{3}.$$

∴ the chance of drawing a *B* is $\tfrac{1}{3}$.

After transferring this (unseen) counter to bag *K*, the chance that it shall contain

$$3\ W, \qquad\qquad \text{ is } \tfrac{2}{3}\times\tfrac{1}{4}; \qquad \text{ i. e. } \tfrac{1}{6}.$$
$$2\ W, \text{ and } 1\ B, \text{ is } \tfrac{2}{3}\times\tfrac{1}{2} + \tfrac{1}{3}\times\tfrac{1}{4}; \text{ i. e. } \tfrac{5}{12}.$$
$$1\ W, \qquad 2\ B, \text{ is } \tfrac{2}{3}\times\tfrac{1}{4} + \tfrac{1}{3}\times\tfrac{1}{2}; \text{ i. e. } \tfrac{1}{3}.$$
$$3\ B, \qquad\qquad \text{ is } \tfrac{1}{3}\times\tfrac{1}{4}; \qquad \text{ i. e. } \tfrac{1}{12};$$

∴ the chance of drawing a *W* from it is

$$\tfrac{1}{6}\times 1 + \tfrac{5}{12}\times\tfrac{2}{3} + \tfrac{1}{3}\times\tfrac{1}{3}; \qquad \text{ i. e. } \tfrac{5}{9}.$$

∴ the chance of drawing a *B* is $\tfrac{4}{9}$.

Before transferring this to bag *H*, the chance that bag *H* shall contain

$$2\ W, \qquad \text{ is } \tfrac{1}{4}\times 1 + \tfrac{1}{2}\times\tfrac{1}{3}; \text{ i. e. } \tfrac{5}{12}.$$
$$1\ W,\ 1\ B, \quad \tfrac{1}{2}\times\tfrac{2}{3} + \tfrac{1}{4}\times\tfrac{2}{3}; \text{ i. e. } \tfrac{1}{2}.$$
$$2\ B, \qquad \tfrac{1}{4}\times\tfrac{1}{3}; \qquad \text{ i. e. } \tfrac{1}{12}.$$

∴, after transferring it, the chance that bag *H* shall contain

$$3\ W, \qquad \text{ is } \tfrac{5}{12}\times\tfrac{5}{9}; \qquad \text{ i. e. } \tfrac{25}{108}.$$
$$2\ W,\ 1\ B, \quad \tfrac{5}{12}\times\tfrac{4}{9} + \tfrac{1}{2}\times\tfrac{5}{9}; \text{ i. e. } \tfrac{50}{108}.$$
$$1\ W,\ 2\ B, \quad \tfrac{1}{2}\times\tfrac{4}{9} + \tfrac{1}{12}\times\tfrac{5}{9}; \text{ i. e. } \tfrac{29}{108}.$$
$$3\ B, \qquad \tfrac{1}{12}\times\tfrac{4}{9}; \qquad \text{ i. e. } \tfrac{4}{108}.$$

Hence the chance of drawing a *W* is

$$\tfrac{1}{108}\times\{25\times 1 + 50\times\tfrac{2}{3} + 29\times\tfrac{1}{3}\};\ \text{ i. e. } \tfrac{17}{27}.$$

i. e. the odds are **17** to **10** on its happening.

Q. E. F.

Dodgson described his final Pillow-Problem (Question 72) as an example of 'transcendental probability'. He went into no details as to why it should be so described, and many people have taken the problem to be a bit of 'leg-pulling'. The problem is as follows:

Question 72. A bag contains 2 counters, as to which nothing is known except that each is either black or white. Ascertain their colours without taking them out of the bag.

Dodgson's answer was:

One is black, and the other white.

His solution, as it appeared in *Pillow-Problems*, was this:

We know that, if a bag contained 3 counters, 2 being black and one white, the chance of drawing a black one would be $\frac{2}{3}$; and that any *other* state of things would *not* give this chance.

Now the chances, that the given bag contains (α) BB, (β) BW, (γ) WW, are respectively $\frac{1}{4}$, $\frac{1}{2}$, $\frac{1}{4}$.

Add a black counter.

Then the chances, that it contains (α) BBB, (β) BWB, (γ) WWB, are, as before, $\frac{1}{4}$, $\frac{1}{2}$, $\frac{1}{4}$.

Hence the chance, of now drawing a black one,

$$= \tfrac{1}{4}\cdot 1 + \tfrac{1}{2}\cdot\tfrac{2}{3} + \tfrac{1}{4}\cdot\tfrac{1}{3} = \tfrac{2}{3}.$$

Hence the bag now contains BBW (since any other state of things would not give this chance).

Hence, before the black counter was added, it contained BW, i.e. one black counter and one white.

Q. E. F.

The flaw in this argument concerns the addition of the black counter, which changes the probabilities. We can eliminate *WWW* because at least one black counter must be in the bag, giving seven possibilities for selecting a counter:

WWB, WBW, WBB, BWW, BWB, BBW, and *BBB.*

The correct probabilities are therefore

$$(\alpha)\tfrac{1}{7}, \quad (\beta)\tfrac{3}{7}, \quad \text{and} \quad (\gamma)\tfrac{3}{7}.$$

So Dodgson's given conclusion was wrong (probably by design), and there is no solution.

A discussion of some of Dodgson's other probability problems, including two other Pillow-Problems and some problems from *The Educational Times*, appears in Chapter 7.

The Dodgson/Price archive

Some years ago, a collection of manuscript puzzles came to light which from 1853 had been shared between Dodgson and his mathematical mentor, Bartholomew Price, Sedleian Professor of Natural Philosophy and a Fellow of Pembroke College.[12] This archive covers a variety of topics, some being well-known puzzles of the day with others invented by Dodgson.

One of these manuscript puzzles concerned probability.

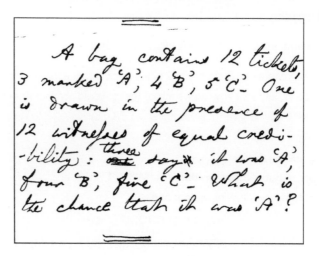

A bag contains 12 tickets, 3 marked 'A', 4 'B', 5 'C'. One is drawn in the presence of 12 witnesses of equal credibility: three say it was 'A', four 'B', five 'C'. What is the chance that it was 'A'?

A probability problem involving twelve tickets

[Courtesy of Edward Wakeling]

Dodgson provided no solution, but an answer is as follows. Let a be the credibility of a witness telling the truth, and $1-a$ of telling a lie. Then if it was A, 3 tell the truth and 9 lie, so the credibility is 3 in 12, or 1 in 4. So the chance that it was A (and no other) is

$$(\tfrac{3}{12} \times \tfrac{1}{4}) + (\tfrac{4}{12} \times \tfrac{3}{4}) + (\tfrac{5}{12} \times \tfrac{3}{4}) = \tfrac{5}{8}.$$

Another puzzle concerned time, and was typed by Dodgson using the Hammond typewriter that he had acquired in May 1888:

A clock face has all the hours indicated by the same mark, and both hands the same in length and form. It is opposite to a looking-glass. Find the time between 6 and 7 when the time as read direct and in the looking-glass shall be the same.

Again, Dodgson gave no solution. At 6 o'clock the minute hand and the hour hand are exactly 180° apart. If the hour hand has moved through $x°$, then the minute hand has moved through $(180 - x)°$. But the hour hand moves at 30° per hour, the minute hand moves at 360° per hour, and the times that have elapsed are identical. It follows that

$$\left(180 - x\right) / 360 = x / 30, \quad \text{so} \quad x = {}^{180}\!/_{13}.$$

This angle represents ${}^{180}\!/_{13} \times {}^{3}\!/_{7} \times 60 = 2\,{}^{4}\!/_{13}$ minutes, so the required time is $30 - 2\,{}^{4}\!/_{13} = 27\,{}^{9}\!/_{13}$ minutes past 6 o'clock.

The monkey and the weight

One particular puzzle from the archive that was circulating at the time was 'The monkey and the weight'. To Dodgson's amusement this hypothetical problem elicited different answers from the various mathematicians and scientists of his day. It is sometimes referred to as 'Lewis Carroll's monkey puzzle', but whether he invented it is open to question. The puzzle is as follows:

A weightless and perfectly flexible rope is hung over a weightless, frictionless pulley attached to the roof of a building. At one end of the rope is a weight which exactly counterbalances a monkey at the other end. If the monkey begins to climb, what will happen to the weight?

Dodgson presented his own solution to Price in a letter dated 19 December 1893; his conclusion was correct.[13]

The monkey and the weight, from Sam Loyd's *Cyclopedia of Puzzles*

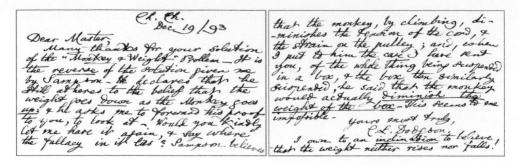

Dodgson's letter to Bartholomew Price on 'The Monkey and the Weight'

[Courtesy of Edward Wakeling]

Four brothers and family monkey

Many puzzles are mentioned by Dodgson in his diaries and letters, but not all were his own invention. Some were popular at the time, such as 'Four brothers and family monkey'. It appeared in various guises, but a version in the hand of Sir John Evans was among the Bartholomew Price mathematical papers cited above. The text is as follows:

The eldest brother comes into a room, finds a heap of nuts on the table, gives one to the monkey, divides the remainder into four equal parts, takes away with him one part and leaves the three other parts. The second brother comes in, finds these three parts, gives one nut to the monkey, divides the remainder into four equal parts of which he takes one, and leaves the other three. The third brother comes in and does likewise, and then the fourth who also does likewise, and the lot that he leaves is still divided into four equal parts. No nuts are cracked. What is the least number of nuts to satisfy these conditions?

Dodgson's typed solution, which he sent to Price, is also among these papers. It is undated, but the use of Dodgson's typewriter again suggests a date of 1888; Evans's letter to Price is dated October 1888.

The origin of this problem is not stated. David Singmaster[14] has cited *Trattato d'Aritmetica*, attributed to Paolo del l'Abbaco (*c*.1370), as one of the earliest versions he could find, but states that a similar problem had been posed in an Armenian manuscript from around the year 640. Whatever its origin, this problem has a long history with many modifications. Dodgson's solution was incorrect because he considered the actions of only three brothers, instead of four, to obtain his answer. Evans thanked Price for his solution (which has not come to light), and offered his own solution of 1789 nuts, which does indeed satisfy the conditions of the problem. However, it is not the smallest solution. This is 1021, and was supplied by Henry Dudeney in *The Canterbury Puzzles* of 1919.[15]

Dodgson publishes his puzzles

Many of Dodgson's invented puzzles are to be found in his letters to correspondents, but they rarely come with a solution. Most of these have now been published with solutions provided. The remarkable fact is that he hardly ever repeated a puzzle. It is evident that he was frequently inventing new ones, and was able to draw on a constant stream of examples. He made copies, probably in a notebook (now lost), with the long-term intention of publishing a puzzle book (which came to nothing).

For the Hull family (Agnes, Alice, Charles, Eveline, and Jessie), Dodgson began constructing a puzzle-book for them to keep, based on examples he had tried with them. As he wrote in his diary on 17 September 1878:[16]

Began a little book of (personal) riddles, for Agnes Hull, "Remarks on the Victims"…

Five days later he recorded:

The idea occurred to me of printing (for private circulation) the above little book, adding a few other acrostics etc. not suited for the general public. In that case I should like a portrait of Agnes as the frontispiece: I must try to photograph her and have it reproduced in Woodbury-type — which would be worth while, as I should probably print 500, or even 1000, copies.

It was not to be. As he noted on 23 August 1879:

Agnes carried off the little MS. book "Remarks on the Victims," and probably dropped it in the road: at all events, it is lost.

Many of these puzzles were probably 'word-related', such as acrostics and poems hiding names, while others may have been numerical or logical.

Dodgson did something similar for the children (Maud, Gwendolen, James, and William) of Robert Gascoyne-Cecil, 3rd Marquess of Salisbury, who was installed as Chancellor of Oxford University in June 1870. He collected together a number of puzzles that he had invented for them and submitted these to Margaret Gatty, editor of *Aunt Judy's Magazine*. Seven puzzles entitled 'Puzzles from Wonderland', all couched in poetic form, were published in December 1870. *Aunt Judy's Magazine* then published 'Solutions' in January 1871 by 'Eadgyth', who is not identified, but does not appear to be Dodgson.

Cats and rats

A lesser-known fact about Dodgson is that he was a magazine columnist on three occasions, contributing puzzles. In the first case he provided a series of mathematical problems that were published in *The Monthly Packet*, a Christian magazine for young ladies edited by Charlotte Yonge. They were initially called 'Spider subjects' and the first (not due to Dodgson) was entitled 'Cats and rats', published in September 1879:

If 6 cats kill 6 rats in 6 minutes, how many will be needed to kill 100 rats in 50 minutes?

The published solution was based on the 'rule of three' and appeared in the December issue. It was followed by 'The cats and rats again' in February 1880, this time by Dodgson, with solutions included in the same issue:

If a cat can kill a rat a minute, how long would it be killing 60,000 rats?

Dodgson added:[17]

My private opinion is, that the rats would kill the cat.

A Tangled Tale

Dodgson's next contributions to *The Monthly Packet* took the form of problems built around a story. The first was called *Romantic Problems, A Tangled Tale, Knot I, Excelsior*, and was published in April 1880. The *Excelsior* problem, illustrated at the beginning of this chapter, concerns speeds of travelling:

Two travellers spend from 3 o'clock till 9 in walking along a level road, up a hill, and home again: their pace on the level being 4 miles an hour, up hill 3, and down hill 6. Find the distance walked: also (within half an hour) time of reaching top of hill.

The answer that Dodgson gave was '24 miles: half-past 6', immediately followed by his full solution:

A level mile takes ¼ of an hour, up hill ⅓, down hill ⅙. Hence to go and return over the same mile, whether on the level or on the hill-side, takes ½ an hour. Hence in 6 hours they went 12 miles out and 12 back. If the 12 miles out had been nearly all level, they would have taken a little over 3 hours; if nearly all up hill, a little under 4. Hence 3½ hours must be within ½ an hour of the time taken in reaching the peak; thus, as they started at 3, they got there within ½ an hour of ½ past 6.

This solution, with Dodgson's review of the twenty-seven answers received, was published in June 1880. There were nine correct solutions, sixteen partially correct, and two were wrong. Dodgson named names, but many were identified as initials or pseudonyms such as 'Sea-Breeze', 'Simple Susan', and 'Money-Spinner' and appeared in a 'Class list'.

```
              CLASS LIST
                   I
    A MARLBOROUGH BOY.        PUTNEY WALKER.
                   II
    BLITHE.                   ROSE.
    E. W.                     SEA-BREEZE.
    L. B.                     SIMPLE SUSAN.
    O. V. L.                  MONEY-SPINNER.
```

Class list for Knot I, *Excelsior*

Over the ensuing months a series of ten 'knots' was published, followed by solutions and a review of the contributions.[18] The last knot, 'Chelsea buns', appeared in November 1884, with the solutions given in the March 1885 issue. Some of the knots contained more than one mathematical puzzle, and some of these required more than an elementary knowledge of arithmetic; two problems involved pre-decimal British money, with twelve pence (*d.*) in a shilling (*s.*) and twenty shillings in a pound (£). The problems in the last nine knots, as summarized by Dodgson, are listed here; his answers (but not his full solutions) are given at the end.

Problems from *A Tangled Tale*

Knot I, *Excelsior*, is given above.

Knot II, *Eligible Apartments* features two problems concealed in a story: one problem is about a dinner party and the other concerns the numbering of houses around a square.

Problem I. The Governor of Kgovjni wants to give a very small dinner party, and invites his father's brother-in-law, his brother's father-in-law, his father-in-law's brother, and his brother-in-law's father. Find the number of guests.

Problem 2. A Square has 20 doors on each side, which contains 21 equal parts. They are numbered all round, beginning at one corner. From which of the four, Nos. 9, 25, 52, 73, is the sum of the distances, to the other three, least?

Knot III, *Mad Mathesis*, concerns two travellers on a circular railway.

Problem. (1) Two travellers, starting at the same time, went opposite ways round a circular railway. Trains start each way every 15 minutes, the easterly ones going round in 3 hours, the westerly in 2. How many trains did each meet on the way, not counting trains met at the terminus itself?

(2) They went round, as before, each traveller counting as 'one' the train containing the other traveller. How many did each meet?

(Continued)

Knot IV, *The Dead Reckoning*, involves the weighing of sacks during a sea voyage.

Problem. There are 5 sacks, of which Nos. 1, 2, weigh 12 lbs.; Nos. 2, 3, 13½ lbs., Nos. 3, 4, 11½ lbs., Nos. 4, 5, 8 lbs., Nos. 1, 3, 5, 16 lbs. Required the weight of each sack.

Knot V, *Oughts and Crosses*, concerns the assignment of up to three marks (× or ○) to each of ten pictures in the Royal Academy.

Problem. To mark pictures, giving 3 ×'s to 2 or 3, 2 to 4 or 5, and 1 to 9 or 10; also giving 3 ○'s to 1 or 2, 2 to 3 or 4 and 1 to 8 or 9; so as to mark the smallest possible number of pictures, and to give them the largest possible number of marks.

Knot VI, *Her Radiancy*, features two problems, one about money and the other on comparing scarves.

Problem 1. A and B begin the year with only £1,000 a-piece. They borrowed nought; they stole nought. On the next New-Year's Day they had £60,000 between them. How did they do it?

Problem 2. L makes 5 scarves, while M makes 2: Z makes 7 while L makes 3. Five scarves of Z's weigh one of L's; 5 of M's weigh 3 of Z's. One of M's is as warm as 4 of Z's: and one of L's as warm as 3 of M's. Which is best, giving equal weight in the result to rapidity of work, lightness, and warmth?

Knot VII, *Petty Cash*, includes an algebra problem about lemonade, sandwiches, and biscuits.

Problem. Given that one glass of lemonade, 3 sandwiches, and 7 biscuits, cost 1s. 2d; and that one glass of lemonade, 4 sandwiches, and 10 biscuits, cost 1s. 5d.: find the cost of (1) a glass of lemonade, a sandwich, and a biscuit; and (2) 2 glasses of lemonade, 3 sandwiches, and 5 biscuits.

Knot VIII, *De Omnibus Rebus*, presents two problems: one is about pigs in sties and the other, entitled 'The Grurmstipths', is on the timing of omnibuses.

Problem 1. Place twenty-four pigs in four sties so that, as you go round and round, you may always find the number in any sty nearer to ten than the number in the last.

Problem 2. Omnibuses start from a certain point, both ways, every 15 minutes. A traveller, starting on foot along with one of them, meets one in 12½ minutes: when will he be overtaken by one?

Knot IX, *A Serpent with Corners*, includes three problems, two on the displacement of water and the other about an oblong garden.

Problem 1. Lardner states that a solid, immersed in a fluid, displaces an amount equal to itself in bulk. How can this be true of a small bucket floating in a larger one?

Problem 2. Balbus states that if a certain solid be immersed in a certain vessel of water, the water will rise through a series of distances, two inches, one inch, half an inch, &c., which series has no end. He concludes that the water will rise without limit. Is this true?

Problem 3. An oblong garden, half a yard longer than wide, consists entirely of a gravel-walk, spirally arranged, a yard wide and 3,630 yards long. Find the dimensions of the garden.

Knot X, *Chelsea Buns*, features two problems about the injuries sustained by Chelsea pensioners and the ages of three sons; between these was another problem, similar to 'Where does the day begin?', which was not answered by the author.

Problem 1. If 70 per cent. have lost an eye, 75 per cent. an ear, 80 per cent. an arm, 85 per cent. a leg: what percentage, at least, must have lost all four?

Problem 3. At first, two of the ages are together equal to the third. A few years afterwards, two of them are together double of the third. When the number of years since the first occasion is two-thirds of the sum of the ages on that occasion, one age is 21. What are the other two?

Answers

Knot II, *Problem 1*: One; the answer to *Problem 2* is given below.

Knot III, (1): 19; (2): The easterly traveller met 12; the other 8.

Knot IV: 5½, 6½, 7, 4½, 3½.

Knot V: 10 pictures; 29 marks; arranged thus:—

```
×   ×   ×   ×   ×   ×   ×   ×   ×   ○

×   ×   ×   ×   ×           ○   ○   ○   ○

×   ×   ○   ○   ○   ○   ○   ○   ○   ○
```

Knot VI, *Problem 1*: They went that day to the Bank of England. *A* stood in front of it, while *B* went round and stood behind it; *Problem 2*: The order is *M, L, Z*.

Knot VII, (1): 8d.; (2): 1s. 7d.

Knot VIII, *Problem 1*: Place 8 pigs in the first sty, 10 in the second, nothing in the third, and 6 in the fourth. 10 is nearer ten than 8; nothing is nearer ten than 10; 6 is nearer ten than nothing; and 8 is nearer ten than 6; *Problem 2*: In 6¼ minutes.

Knot IX, *Problems 1 and 2*: Detailed solutions, but no Answers, are given; *Problem 3*: 60, 60½.

Knot X, *Problem 1*: Ten; *Problem 3*: 15 and 18.

As an example of Dodgson's discussion of a *Tangled Tale* problem, let us look in greater detail at Knot II, *Problem 2*: entitled 'The Lodgings', it requires the use of the Pythagorean theorem on right-angled triangles. It concerns a tutor and his two scholars, Hugh and Lambert, who are looking for lodgings. They come across a square in the town that has four houses displaying rooms for rent, but all are single rooms. As the problem states, there are twenty houses on each side of the square with doors dividing the side into twenty-one equal parts; the available lodgings are at numbers 9, 25, 52, and 73. The tutor decides to make one room a 'day-room' and take the rest as bedrooms. He set this problem for his scholars:

"See, boys! Twenty doors on a side! What symmetry! Each side divided into twenty-one equal parts! It's delicious!…We will take as our day-room the one that gives us least walking to do to get to it."

"Must we walk from door to door, and count the steps?" said Lambert.

"No, no! Figure it out, my boys, figure it out!"

So which house should be used for the day-room? Dodgson's solution was:

Let A be No. 9, B No. 25, C No. 52, and D No. 73.

Then $AB = \sqrt{(12^2 + 5^2)} = \sqrt{169} = 13$;

$AC = 21$;

$AD = \sqrt{(9^2 + 8^2)} = \sqrt{145} = 12 +$

(N.B. *i.e.* "between 12 and 13.")

$BC = \sqrt{(16^2 + 12^2)} = \sqrt{400} = 20$;

$BD = \sqrt{(3^2 + 21^2)} = \sqrt{450} = 21 +$;

$CD = \sqrt{(9^2 + 13^2)} = \sqrt{250} = 15 +$;

Hence sum of distances from A is between 46 and 47; from B, between 54 and 55; from C, between 56 and 57; from D, between 48 and 51. (Why not "between 48 and 49"? Make this out for yourselves.) Hence the sum is least for A.

In 1884 Dodgson considered publishing the ten Knots as a separate book called *A Tangled Tale* with illustrations by Arthur Burdett Frost, who had illustrated Dodgson's book of humorous poetry entitled *Rhyme? and Reason?* in 1883. On 3 July 1885 Dodgson noted in his diary:[19]

Wrote to Macmillan and Swain [the engraver] about *Tangled Tale*, which I hope to get out as a Christmas book this year.

The book was published in towards the end of December 1885, probably too late to be a Christmas book. Dodgson noted that he received his copy on 22 December. It was to become a favourite of Josiah Willard Gibbs, the applied mathematician and physical chemist who was later praised by Albert Einstein as 'the greatest mind in American history'.

Problems involving algebra

Several of the mathematical questions in *Pillow-Problems* are algebraic in nature, requiring some use of common sense in interpreting the answer. An example is as follows:

Question 8. Some men sat in a circle, so that each had 2 neighbours; and each had a certain number of shillings. The first had 1/ [one shilling] more than the second, who had 1/ more than the third, and so on. The first gave 1/ to the second, who gave 2/ to the third, and so on, each giving 1/ more than he received, as long as possible. There were then 2 neighbours, one of whom had 4 times as much as the other. How many men were there? And how much had the poorest man at first?

Dodgson's solution was as follows:

Let m = No. of men, k = No. of shillings possess by the last (i.e. the poorest) man. After one circuit, each is a shilling poorer, and the moving heap contains m shillings. Hence, after k circuits, each is k shillings poorer, the last man now having nothing, and the moving heap contains mk shillings. Hence the thing ends when the last man is again called on to hand on the heap, which then contains $(mk + m - 1)$ shillings, the penultimate man now having nothing, and the first man having $(m - 2)$ shillings.

It is evident that the first and last man are the only 2 neighbours whose possessions can be in the ratio '4 to 1'. Hence either

$$mk + m - 1 = 4(m - 2),$$
$$\text{or else} \quad 4(mk + m - 1) = m - 2.$$

The first equation gives $mk = 3m - 7$, i.e. $k = 3 - 7/m$, which evidently gives no integral values other than $m = 7$, $k = 2$.

The second gives $4mk = 2 - 3m$, which evidently gives no positive integral values.

Hence the answer is '7 men; 2 shillings'.

Making sense of an answer was one trait that Dodgson was keen to instil in his undergraduates, and he used various problems to make this point clear. One such problem involved the 'rule of three' applied to proportion and inverse proportion. Reminiscent of 'Cats and rats', it went as follows:

If it takes 10 men so many days to construct a wall, how long will it take 300,000 men to build a similar wall.

Assuming the working day to be 24 hours, undergraduates would often calculate the time using the inverse rule of three and get 2.88 seconds! As Dodgson indicated, such an answer makes no sense of the situation since the majority of the men could get nowhere near the wall. This was an important feature of mathematics that was usually lost when work was taken mainly from textbook exercises, as Dodgson knew well from his own schooldays.

This next example is previously unpublished and is one of the questions that Dodgson devised and wrote out with the probable intention of publishing them in a book of recreational mathematics.[20] The manuscript is dated 26 February 1883 and concerns two brothers wishing to divide a legacy between them. It appears here with its solution up the left-hand side.

Manuscript of the legacy problem

[Morris L. Parrish Collection, Princeton University]

Problems involving geometry

We next set out a number of problems that Dodgson presented to his child-friends involving ideas from geometry.

A geometrical paradox

The following puzzle, which seems to have been invented by O. X. Schlömilch in 1868, was found among Dodgson's papers after his death:

Start with an 8 × 8 grid of 64 squares and cut it into four pieces, as shown below. If we rearrange the pieces, we obtain a 5 × 13 grid of 65 squares. Where did the extra square come from?

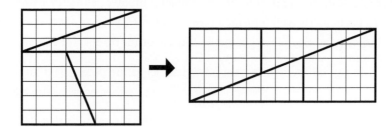

This paradox can also be extended to larger grids of squares; for example, a 21 × 21 grid of 441 squares can be rearranged to give a 13 × 34 grid of 442 squares. In each case the paradox is removed by careful drawing, which shows that there is a very thin gap with area 1 in the middle of the rearranged grid.

In his diary for 16 February 1890 we learn how Dodgson learned of the problem from Henry Giffard, a former Junior Mathematical Scholar at Oxford:[21]

Giffard gave me the old puzzle of cutting up a chequered board into four pieces, so as to look as if a square were gained.

Areas being a^2 and (apparently) $2a^2 - 3ax + x^2$, which has to be $= a^2 + 1$.

Values for a, x, may be taken thus

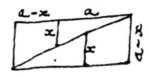

a	3	8	21	55	144	377	etc.
x	1	3	8	21	55	144	etc.

where series 1, 3, 8, etc. is found by formula

$$-\underline{1} + 3.\underline{3} = \underline{8} \qquad -\underline{3} + 3.\underline{8} = \underline{21} \quad \text{etc}$$

The numbers 5, 8, and 13 that appear in the original problem are all *Fibonacci numbers*, a sequence that begins

$$1, 1, 2, 3, 5, 8, 13, 21, 34, 55, 89, 144, \ldots,$$

where each successive number is the sum of the previous two. At no stage does Dodgson refer to them by name.

From the above diary entry, and from Dodgson's notebook in the Parrish Collection at Princeton University, we learn that from 1890 to 1893 he continued to extend this seeming paradox to larger grids of squares.

Pillow-Problems revisited

Many of the questions that appear in *Pillow-Problems* involve geometry, a topic on which Dodgson was particularly well informed (see Chapter 2). We present two examples, one from two-dimensional Euclidean geometry and the other from solid geometry.

Question 2. In a given Triangle to place a line parallel to the base, such that the portions of sides, intercepted between it and the base, shall be together equal to the base.

Dodgson's solution is as follows: the requirement in the Question is that $BD + CE = BC$. We notice how he split his solution into the traditional approaches of Analysis (where he assumed that $BD + CE = BC$ and described the construction) and Synthesis (where he reversed the argument).

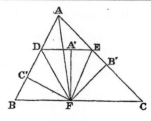

(Analysis.)

Let ABC be the Triangle, and DE the required line, so that $BD + CE = BC$.

From BC cut off BF equal to BD; then $CF = CE$.

Join DF, EF.

Now $\angle BDF = \angle BFD = [\text{by I. 29}] \angle FDE$;

Similarly $\angle CEF = \angle FED$;

∴ $\angle s$ BDE, CED, are bisected by DF, EF, and F is centre of $\odot$ escribed to $\triangle ADE$.

Drop, from F, ⊥s on BD, DE, EC; then these ⊥s are equal.

Hence, if AF be joined, it bisects $\angle A$.

Hence construction.

(Synthesis.)

Bisect $\angle A$ by AF: from F draw FB', FC', ⊥AC, AB: also draw $FA' \perp BC$ and equal to FB': and through A' draw $DE \perp FA'$, i. e. ∥ BC. Then DE shall be line required.

∵ $\angle s$ at A', B', C', are right, and $FA' = FB' = FC'$,

∴ $\angle s$ BDE, CED, are bisected by DF, EF.

Now $\angle BFD = \angle FDA'$; ∴ it $= \angle BDF$; ∴ $BF = BD$;

Similarly $CF = CE$; ∴ $BC = BD + CE$.

Q. E. F.

Question 49. If four equilateral Triangles be made the sides of a square Pyramid: find the ratio which its volume has to that of a Tetrahedron made of the Triangles.

Using basic trigonometry, and taking the side of each triangle to be 1, Dodgson calculated the altitude of the tetrahedron to be $\sqrt{2}/\sqrt{3}$ and its volume to be $\sqrt{2}/12$, whereas the altitude of the pyramid is $\sqrt{2}/2$ and its volume is $\sqrt{2}/6$. So the required ratio is 2.

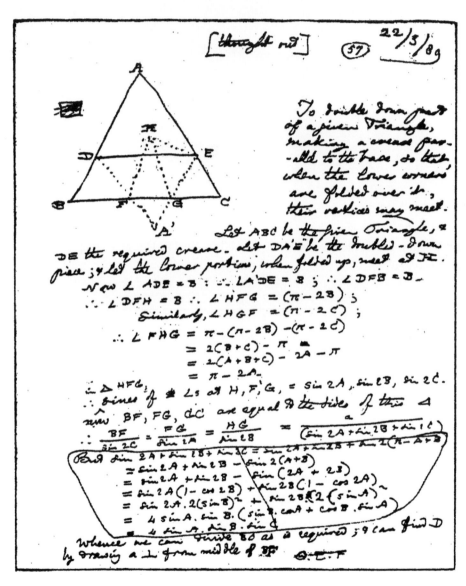

The manuscript of another geometrical 'Pillow-Problem'

[Governing Body of Christ Church, Oxford]

Colouring maps

The *four-colour problem* is a well-known mathematical question on the colouring of maps. It was first posed in 1852, but was not completely answered until 1976:[22]

Can every map be coloured with only four colours in such a way that all pairs of regions sharing a common border are coloured differently?

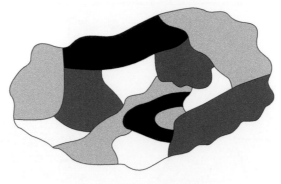

A four-coloured map

As his nephew Stuart Dodgson Collingwood described it, Dodgson adapted it as a game for two players:[23]

Another favourite puzzle was the following – I give it in his own words:–
A is to draw a fictitious map divided into counties.
B is to colour it (or rather mark the counties with *names* of colours) using as few colours as possible. Two adjacent counties must have *different* colours.
A's object is to force B to use as *many* colours as possible.
How many can he force B to use?

Fortunatus's Purse

In *Sylvie and Bruno Concluded*, Dodgson's last story book, there is a scene in which Lady Muriel is taking tea with her father, the Earl of Ainslie, while hemming pocket handkerchiefs. Along comes Mein Herr, a German professor, who reminds her about 'the puzzle of the Paper Ring', and tells her how she can sew together three handkerchiefs to make *Fortunatus's Purse*, an object that supposedly holds all the wealth of the world inside it.

Constructing Fortunatus's purse

The 'paper ring' is a *Möbius band*, an object with just one side and one edge, obtained by joining the two ends of a rectangular strip of paper after giving one end a 180° twist. Fortunatus's Purse is the object we would then obtain if we could join the top and bottom of the paper ring after giving one of them a 180° twist. It does not exist in three dimensions, has no outside and no inside, and is known to mathematicians as a *projective plane*.

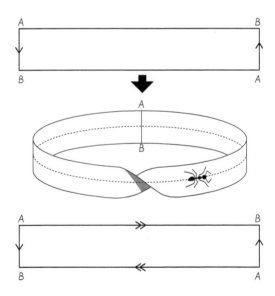

Constructing a Möbius band and a projective plane

Alice's Puzzle-Book

Dodgson devised enough puzzles and games to contemplate compiling a book devoted entirely to these inventions. On 1 March 1875 he recorded in his diary:[24]

Wrote to Tenniel on the subject of an idea, which I first entered in my memorandum book, January 8th, of printing a little book of original puzzles etc. which I think of calling *Alice's Puzzle-Book*. I want him to draw a frontispiece for it. (He consented March 8).

In time the book came to be known as *Original Games and Puzzles*, but it was never published. It is among a list of several books that he was contemplating, as recorded in his diary on 23 November 1881. It was listed again on 29 March 1885, this time described as

A collection of Games and Puzzles of my devising, with fairy-pictures by Miss E. G. Thomson [including a puzzle-picture for which Dodgson drew her a sketch]. This might also contain my "Memoria Technica" for dates etc., my "Cipher-writing," scheme for Letter-registration, etc. etc.

Some sections were even set in type, with many puzzles still in manuscript. On 12 November 1897 (just a few weeks before his death) he wrote in his diary:

An inventive morning! After waking, and before I had finished dressing, I had devised a new, and much neater, form in which to work my Rules for Long Division, and also decided to bring out my "Games and Puzzles," and Part III of *Curiosa Mathematica*, in *Numbers*, in paper covers, paged consecutively, to be ultimately issued in boards.

Dodgson always had several projects in the pipeline and a number of these did not come to fruition following his death in 1898, the puzzle book being one of them.

Memoria Technica

Memoria Technica was a system devised by Dodgson in the mid-1870s for remembering key dates and mathematical data. It was based on a previous scheme dated 1730 by Dr Richard Grey, a Northamptonshire rector, and involved rhyming couplets.[25] To remember a date Dodgson would create a word or phrase, a rhymed couplet, or a short verse, whose *consonants* encode the number to be remembered, using the following table:

1	2	3	4	5	6	7	8	9	0
b	*d*	*t*	*f*	*l*	*s*	*p*	*h*	*n*	*z*
c	*w*	*j*	*qu*	*v*	*x*	*m*	*k*	*g*	*r*

For example, Dodgson's verse for log 2 = 0·3010300... was

> Two jockeys to carry
> Made *that* racer tarry.

Here the final seven consonants, *t r c r t r r*, give the digits 3, 0, 1, 0, 3, 0, 0.

1	2	3	4	5	6	7	8	9	0
b	*d*	*t*	*f*	*l*	*s*	*p*	*h*	*n*	*z*
c	*w*	*j*	*qu*	*v*	*x*	*m*	*k*	*g*	*r*

Each digit is represented by one or other of two consonants, according to the above table: vowels are then inserted *ad libitum* to form words, the significant consonants being always at the *end* of a line: the object of this is to give the important words the best chance of being, by means of the rhyme, remembered accurately.

The consonants have been chosen for the following reasons.

(1) *b, c*, first two consonants.
(2) *d* from "deux"; *w* from "two"
(3) *t* from "trois"; *j* was the last consonant left unappropriated.
(4) *f* from "four"; *qu* from "quatre".
(5) *l* = 50; *v* = 5.
(6) *s, x,* from "six."
(7) *p, m,* from "septem."
(8) *h* from "huit"; *k* from ὀκτώ.
(9) *n* from "nine"; *g* from its shape.
(0) *z, r,* from "zero."

They were also assigned in accordance, as far as possible, with the rule of giving to each digit one consonant in common use, and one rare one.

Since *y* is reckoned as a vowel, many whole words, (such as "ye", "you", "eye"), may be put in to make sense, without interfering with the significant letters.

Take as an example of this system the two dates of "Israelites leave Egypt — 1495," and "Israelites enter Canaan — 1455" :—

"Shout again! We are free!"
Says the loud voice of glee.
"Nestle home like a dove,"
Says the low voice of love.

Ch. Ch.
June 27/77

Dodgson's *Memoria Technica*

[Morris L. Parrish Collection, Princeton University]

Dodgson worked with his physics colleague Robert Baynes on the *Memoria Technica*. On 27/28 October 1875 he recorded in his diary that he:[26]

Sat up until nearly 2 making a "Memoria Technica" for Baynes for logarithms of primes up to 41. I can now calculate in a few minutes almost any logarithm without book.

To do so, he found that just twenty-six logarithms had to be remembered in order to obtain the logarithm (to base 10) of any number correct to seven decimal places: these were the logarithms of the primes up to 20, the numbers from 101 to 109, and those from 1001 to 1009. For example, to remember the logarithm of 106 = 2.0253059 he composed a rhyming couplet that included the phrase: 'I have changed from a dove to a raven'; here the last six consonants *d v t r v n* give the required decimal digits 2, 5, 3, 0, 5, 9.

Dodgson also created other 'phrases' that gave π to 71 decimal places and e to 30 decimal places, and on 21 March 1878 his diary records that he had written verses that gave him π to 102 decimal places.[27] He also recorded that it took him just nine minutes to work out the thirteenth root of 87,654,327, thirteen minutes to calculate the tenth power of 237,541, and fourteen minutes to calculate π^π, all correct to four or five decimal places.

Gray's *Memoria Technica* had originally required the memory of gibberish phrases in order to remember numbers and dates. Dodgson realized that if a person had difficulty in remembering a number, they would have as much difficulty in remembering a gibberish word – hence his improvement requiring only the memory of a short appropriate rhyming couplet. But whether remembering the phrases and verses is more effective than using tables is open to doubt. Dodgson developed the idea in June 1877 and he continued to modify and improve his method over several years.

The Educational Times and Nature

The Educational Times was a monthly London periodical that published mathematical problems and their solutions, beginning in 1862. Dodgson made ten contributions to it between 1879 and his death, with one published posthumously. In November 1879 he sent a paper entitled 'Practical Hints on Teaching. Long Multiplication worked with a single line of Figures'. He wrote to the editor, including a worked example, saying:

if the following brief method of working Long Multiplication should prove to be new, I hope you may think it worth publishing.

The editor, William John Clarke Miller, duly published the method.

Encouraged by this, Dodgson began to respond and contribute to Questions published in *The Educational Times* on a variety of topics: probability (1 May 1885), geometrical probability, infinitesimals, and arithmetic (1 June 1888), algebra (1 February 1889), trigonometry (1 May 1893), number-guessing (1 February 1895), and finding the day of the week for any given date (1 September 1897). His final contribution concerned divisibility by 9 or 11; it appeared on 1 February 1899, a year after his death.

Rules for divisibility

There are several simple rules for testing the divisibility of a given number by certain smaller ones, such as the following.

To test whether a given number is divisible by 3, add its digits and find whether the result is divisible by 3:
for example, 12345 is divisible by 3 because $1 + 2 + 3 + 4 + 5 = 15$ is divisible by 3.

To test whether a number is divisible by 4, find whether the number given by the last two digits is divisible by 4:

for example, 12348 is divisible by 4 because the number 48 is divisible by 4.

Towards the end of his life Dodgson developed an interest in formulating rules for divisibility, and wrote 'Divisibility by Seven' which was published in *Knowledge* in July 1884. There was nothing new in Dodgson's paper, but his subsequent paper 'Brief Method of Dividing a Given Number by 9 or 11' was his own invention. On 27 September 1897 he wrote in his diary:[28]

Dies notandus. Discovered rule for dividing a number by 9, by mere addition and subtraction. I felt sure there must be an analogous one for 11, and found it, and proved first rule by Algebra, after working about nine hours!

This was followed on the next day by:

Dies cretâ notandus! I have actually *superseded* the rules discovered yesterday! My new rules require to ascertain the 9-remainder, and the 11-remainder, which the others did *not* require: but the new ones are much the quickest. I shall send them to *The Educational Times*, with date of discovery.

His methods also extended to division by 13, 17, 19, and 41, and by numbers that were close to a power of 10.

Dodgson's note did not immediately appear in *The Educational Times*, but was first published in *Nature* on 14 October 1897. A further galley proof exists for 'Divisibility by 19, 29 up to 119', dated 1 December, but Dodgson's death in January 1898 prevented this from being published.[29] The paper was probably intended for *Curiosa Mathematica, Part III*, or for his book on 'Games and Puzzles'.

A third note was submitted to *Nature* on 'Abridged Long Division', written on 21 December 1897, just twenty-four days before he died.[30] It was published posthumously on 20 January 1898, and is essentially a generalized form of 'A Brief Method of Dividing a Given Number by 9 or 11'.

Finding the day of the week

Another contribution that Dodgson made to *Nature* in 1887 was 'To Find the Day of the Week for Any Given Date'. This was published on 31 March, several days after he invented it. On 8 March he wrote:[31]

Discovered a Rule (since last Thursday) for finding day of week for any given day of the month. There is less to remember than in any other Rule I have met with.

To try out his method he listed ten examples that he had worked out mentally in under 250 seconds.[32] His method involved four calculations: two for the year, one for the month, and one for the day, and his examples ranged from 31 July 188 (Wednesday) to 18 March 2673 (Tuesday). Dodgson's method is given here.

Finding the day of the week for any given date

Take the given date in 4 portions, viz. the number of centuries, the number of years over, the month, the day of the month.

Compute the following 4 items, adding each, whenever found, to the total of the previous items. When an item or total exceeds 7, divide by 7, and keep the remainder only.

The Century-Item: Divide by 4, take overplus from 3, multiply remainder by 2.

The Year-Item: Add together the number of dozens, the overplus, and the number of 4's in the overplus.

The Month-Item: If it begins or ends with a vowel, subtract the number, denoting its place in the year, from 10. This, plus its number of days, gives the item for the following month. The item for January is '0'; for February or March (the third month), '3'; for December (the 12th month), '12.'

The Day-Item: is the day of the month.

The total, thus reached, must be corrected, by deducting '1' (first adding 7, if the total be '0'), if the date be January or February in a Leap Year: remembering that every year, divisible by 4, is a Leap Year, excepting only the century-years, in New Style, when the number of centuries is *not* so divisible (e.g. 1800).

The final result gives the day of the week, '0' meaning Sunday, '1' Monday, and so on.

EXAMPLE

1783, September 18

17, divided by 4, leaves "1" over; 1 from 3 gives "2"; twice 2 is "4."

83 is 6 dozen and 11, giving 17; plus 2 gives 19, *i.e.* (dividing by 7) "5." Total 9, i.e. "2."

The item for August is "8 from 10," i.e. "2"; so for September, it is "2 plus 3," *i.e.* "5." Total 7, i.e. "0," which goes out.

18 gives "4." Answer, "*Thursday.*"

Number-guessing and other puzzles

In the 1890s Dodgson became interested in number-guessing problems. Here a person thinks of a number, and after a series of questions and arithmetical calculations the original number can be deduced. On 6 January 1895 he recorded in his diary:[33]

Out walking, yesterday, I invented a "number-guessing" puzzle, which contains, four times, "divide by 3: name remainder."

No surviving manuscript of this game has come to light, but it was one of many such puzzles that Dodgson invented. On 3 February 1896 he wrote in his diary:

Have been for some days devising an original kind of Number-Guessing Puzzle giving *choice* of numbers to operate with, and have this morning brought it to a very satisfactory form.

Dodgson wrote out the puzzle as a series of questions, but within a few days (on 6 February) he had modified the text.[34] The only intermediate answer that the enquirer needed to know was whether the result was odd or even, and how many times the final answer was divisible by 7.[35]

Number-guessing

A. "Think of a number."

B. [thinks of 23]

A. "Multiply by 3. Is the result odd or even?"

B. [obtains 69] "It is odd."

A. "Add 5, or 9, whichever you like."

B. [adds 9, & obtains 78]

A. "Divide by 2, & add 1."

B. [obtains 40]

A. "Multiply by 3. Is the result odd or even?"

B. [obtains 120] "It is even."

A. "Subtract 2, or 6, whichever you like."

B. [subtracts 6, and obtains 114]

A. "Divide by 2, & add 29, or 38, or 47, whichever you like."

B. [adds 38, & obtains 95]

A. "Add 19 to the original number, and tack on any figure you like."

B. [tacks on 5, & obtains 425]

A. "Add the previous result."

B. [obtains 520]

A. "Divide by 7, neglecting remainder."

B. [obtains 74]

A. "Again divide by 7. How often does it go?"

B. "Ten times."

A. "The number you thought of was 23."

How did A know? In order to find B's original choice, A multiplies B's final answer by 4 and subtracts 15; then if B's first answer was 'even', subtract 3 more, and if B's second answer was 'even' subtract 2 more; here, $(10 \times 4) - 15 - 2 = 23$. In fact, Carroll slightly miscalculated when constructing this puzzle, for if B starts with 4, 6, or 8, then the answers are 'even', 'even' and 'seven times', so A cannot know B's original number. But the matter is easily rectified by changing 'or 38, or 47' to 'or 33, or 37' in A's seventh instruction, and all then works out correctly.

Dodgson added a note to his diary on 15 February:

My number-guessing puzzle is *not* insoluble. Adamson [a Fellow of St. John's College] had made out the principle, and guessed one himself.

According to his diary he tried out his number-guessing puzzle with the children of a fellow colleague at Christ Church in March 1897, so it remained part of his repertoire.

Two other puzzles

As we have seen, Dodgson was frequently called upon to entertain groups of young people as a favour to friends. He could capture the attention of large groups with his puzzles and games. His diary entry for 6 October 1897 indicates the kind of entertainment he provided:[36]

I gave them the "Mr. C. and Mr. T." and "M.O.W.S." stories – the saying a lot of figures by *Memoria Technica* – 142857 – and the reversed £. s. d.

His stories were illustrated by drawings made as the adventure unfolded, often resulting in a puzzle picture of an animal.

The '142857' puzzle was a simple arithmetical curiosity, popular in Victorian times, which Dodgson enjoyed showing to his child-friends. He asked them to multiply this magic number by 2, 3, 4, 5, 6, and 7:

Begin at the '1' in each line and it will be the same order of figures as the magic number up to six times that number, while seven times the magic number results in a row of 9's.

A Magic Number: 142857

285714   twice that number.
428571   thrice that number.
571428   four times that number.
714285   five times that number.
857142   six times that number.
999999   seven times that number.

The reason for this arises from the fact that the first six of these numbers appear in the cyclical decimal representation of the fraction $\frac{1}{7} = \frac{142857}{999999} = 0.142857142857142\ldots.$[37]

The 'reversed £. s. d.' was a money puzzle, described by Dodgson as follows;[38] recall that in pre-decimal coinage there were twelve pence (*d.*) in a shilling (*s.*) and twenty shillings in a pound (£):

Put down any number of pounds not more than twelve, any number of shillings under twenty, and any number of pence under twelve. Under the pounds put the number of pence, under the shillings the number of shillings, and under the pence the number of pounds, thus reversing the line.

Subtract.

Reverse the line again.

Add.

Answer, £12 18s. 11d., *whatever* numbers may have been selected.

For example, suppose that you write down	£3 14s 9d.
Reversing this gives	£9 14s 3d.
Subtracting the smaller amount from the larger gives	£5 19s 6d.
Reversing this gives	£6 19s 5d.
Adding these last two amounts gives	**£12 18s 11d.**

The reason that this works is that if the original amount is £x ys. zd. (where $x > z$), then reversing this gives £z ys. xd. Subtracting the latter from the former gives

$$£(x - z - 1)\ 19s.\ (12 + z - x)d.$$

and reversing this gives

$$£(12 + z - x)\ 19s.\ (x - z - 1)d.$$

Finally, adding these last two amounts gives £12 18s. 11d.

Conclusion

It is difficult to quantify all the puzzles and games that Dodgson devised because they are still being discovered. A simple riddle told to his child-friend Mabel Burton was recently discovered among her recorded reminiscences of Dodgson, together with all her letters and presentation copies. They had remained in the family for three generations and were not made public until a recent inheritor decided to sell everything. The riddle of 'the weight of the fish' that she remembers hearing from Dodgson was:

If a fish weighed six pounds and half its own weight, what was the total weight of the fish?

The answer is twelve pounds.

As we noted above, Dodgson rarely repeated a puzzle, such was the fund of ideas that flowed from his mind: no other child has recorded this particular puzzle. Several collections of Dodgson's puzzles have now been published (see the Further reading), but others await to be made public.

Charles Lutwidge Dodgson, aged 42

[By kind permission of the Dodgson family]

CHAPTER 7

Mathematical legacy

FRANCINE F. ABELES

In the preface to his book *Lewis Carroll in Numberland*, Robin Wilson asked three questions about Dodgson's mathematics: What mathematics did he do? How good a mathematician was he? How influential was his work? Here we build on the material of the previous chapters with the goal of answering these questions and a few more.

This chapter is organized into seven sections: geometry, trigonometry, algebra, logic, voting, probability, and cryptology. In these sections we examine the state of the subject during Dodgson's time, the important mathematical ideas and methods that he created or developed, the mathematicians that he worked with or influenced, his relevant publications and those by other mathematicians who may have influenced him, the mathematical ideas and methods that his work foreshadowed in the 20th century, and some modern mathematicians who have been inspired by his work.[1]

Geometry

Public perception of Dodgson's geometrical work came from the Dover republication in 1973 of the second (1885) edition of *Euclid and His Modern Rivals* (see Chapter 2). That volume dealt primarily with his work as the mathematics lecturer at Christ

The Mathematical World of Charles L. Dodgson (Lewis Carroll). Robin Wilson and Amirouche Moktefi.
Oxford University Press (2019). © Oxford University Press 2019.
DOI: 10.1093/oso/9780198817000.001.0001

Two editions of *Euclid and his Modern Rivals*

Church, and with his firm belief in the value of Euclid as the main vehicle for teaching geometry to undergraduates. How different the perception of Dodgson as a geometer might have been had Dover also chosen to publish his *Curiosa Mathematica, Part I. A New Theory of Parallels*, which appeared nine years after the first appearance of *Euclid and his Modern Rivals*. Here we see Dodgson as a mature mathematician, fully aware of the non-Euclidean geometries that had been discovered earlier in the century.

As a deeply religious man Dodgson considered his mathematical abilities to be a gift that he should use in the service of God, and he linked his work in geometry with his religious beliefs through the way he perceived natural theology and the nature of mathematical truth. In his time mathematics was considered uniquely capable of generating truths from axioms that captured the nature of reality. His need to study and develop logical rules for reasoning reflected this conviction.

The parallel postulate

In his introduction to the 1885 edition of Dodgson's *Euclid and his Modern Rivals*, the eminent British-born Canadian geometer H. S. M. Coxeter wrote, concerning two of Dodgson's tables of results related to the parallel postulate:[2]

Tables I and II are his nearest approach to the subject of non-Euclidean geometry…One is tempted to speculate on what might have happened if Cayley or Clifford had met Dodgson and convinced him that there is a logically consistent "hyperbolic" geometry in which the "absolute" propositions in Table I still hold while all the statements in Table II are false (and the nineteenth proposition fails for any sufficiently large circle).

The 19th proposition that Coxeter referred to here is:

In every circle, the inscribed equilateral hexagon is greater [in area] than any one of the segments that lie outside it.

This was Dodgson's unusual alternative to Euclid's parallel postulate (see Chapter 2). Note that it is a closed form that avoids the difficulty of knowing how parallel lines behave at infinity – whether they remain apart, as in Euclidean geometry, or whether they become infinitesimally close, as in hyperbolic geometry.

Dodgson recognized the existence of both hyperbolic and elliptic geometries. His assessment of hyperbolic geometry was negative – and we will explore his reasons for this – but he rejected elliptic geometry entirely. Adhering to his own interpretation of Euclid's geometry he could not logically accept hyperbolic geometry, and because his idea of an axiom was based on an average person's acceptance without proof, he objected to the tacit appearance of infinities and infinitesimals in Euclidean geometry.

Dodgson was alert to issues about Euclid's geometry appearing in scientific journals. A long article, 'Chats about geometrical measurement', which took the form of a dialogue between two characters A. and M., appeared in the 24 October 1884 issue of *Knowledge*, a weekly popular science journal, written by its editor, Richard Proctor. Two weeks later Dodgson's response to it, entitled 'Euclid's theory of parallels', was published. His response dealt with comments that Proctor had made, which Dodgson faulted from a logical point of view.[3]

Dodgson's attempt to dispense with Euclid's parallel postulate was his reason for writing *A New Theory of Parallels*, first published in July 1888. His aim was to replace Euclid's axiom with one that would be 'intuitively' true – by which he meant an axiom that does *not* involve infinities and infinitesimals. He regarded Euclid's parallel postulate

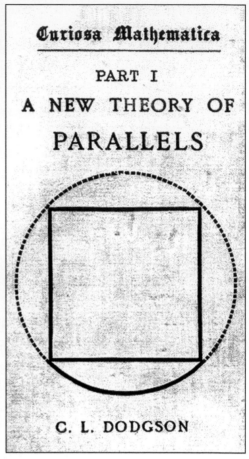

Two editions of *A New Theory of Parallels*

as something that was true for *finite* magnitudes only, and he believed that this is what Euclid really intended but did not state directly.

The second appendix to *A New Theory of Parallels* contains Dodgson's investigations into the role of non-finite processes in the plane, but its title, 'Is Euclid's axiom true?', seems odd since an axiom in mathematics is *always* true. So what did Dodgson intend? The answer to this question involves the defects that Dodgson saw in Euclid's parallel axiom, and the way that he thought about axioms generally. In his *Supplement to Euclid and his Modern Rivals*, published in April 1885, he wrote that an axiom always requests the voluntary assent of readers to some truth for which no proof is offered and which they are not logically compelled to grant, and he went on to say that accepting an axiom depends largely on the amount of intuitive truth that the reader has already grasped in connection with it.[4]

We see here that Dodgson did not subscribe to the idea that axioms are empirical laws subject to verification. Rather, axioms must be acceptable to the faculties of reason which, by its nature, cannot apprehend infinities and infinitesimals. But what exactly allows the reader to apprehend that *truth* in whatever amount? Dodgson did not give his answer here, but did so three years later in his preface to *A New Theory of Parallels* where he justified his own alternative to Euclid's parallel axiom.

The Archimedean axiom

The *Archimedean axiom* for real numbers states that if a and b are positive numbers, then there is a positive integer n for which the product na exceeds b; for example, if $a = 0.1$ and $b = 5$, we can take n to be any integer larger than 50. It follows that the set of real numbers can contain no infinitesimal quantities. Archimedes stated the result in his book *On the Sphere and Cylinder*, and it appears in Euclid's *Elements*, Book V, Definition 4, in the form:

Magnitudes are said to *have a ratio* to one another which are capable, when multiplied, of exceeding one another.

Euclid then assumes a version of this axiom in Book X, Proposition 1:

Two unequal magnitudes being set out, if from the greater there be subtracted a magnitude greater than its half, and from that which is left a magnitude greater than its half, and if this process be repeated continually, there will be some magnitude which will be less than the lesser magnitude set out.

In his diary entries of 24–26 March 1888, Dodgson confided that he could prove that 'no triangle has angle-sum greater than two right-angles' without using Euclid's result.[5] Instead, he used his new axiom which does not depend on the Archimedean axiom.

Dodgson's new axiom, which was equivalent to Euclid's parallel axiom (although he did not recognize it as such), went through many forms before he settled on the one that eventually appeared in *A New Theory of Parallels*, completing the manuscript for the book on 13 April 1888. In Appendix IV of the first two editions, Dodgson included his own failed attempt to *prove* his parallel axiom, calling it a 'Will-o'-the-Wisp'. In doing so he found a curious alternative form for his axiom:[6]

if an equilateral triangle be inscribed in a circle, it cannot be made to lie partly within it, with each vertex outside, and the circle cutting each side in two points.

We take up the issue of proving the parallel axiom further on.

Dodgson firmly believed that Euclid had specifically excluded infinities and infinitesimals, and in his attempts to prove Euclid's parallel axiom he was determined to avoid the Archimedean axiom under which they cannot exist. In other words, allowing parallelism to apply to *infinite* lines introduces not only the necessity of infinitesimals, but also the possibility of there being *another* parallel axiom, a hyperbolic one.

Ordinary human reasoning involves visual processing, and in processing parallel lines we cannot 'see' them to be infinitely long. Dodgson believed that accepting an axiom requires ordinary, rather than mathematical, reasoning. In support of the superiority of his axiom he wrote:[7]

while *every* other Theory (that I have seen), which attempts to supersede Euclid's 12th Axiom, introduces the ideas of Infinities and Infinitesimals, *mine* dispenses *wholly* with their aid, and deals with nothing but what is, by universal consent, absolutely *within* the field of Human Reason.

To his credit, Dodgson understood that hyperbolic geometry is necessarily a non-Archimedean geometry because it admits infinitesimals, and he was one of the few mathematicians of his time to appreciate this fact. However, in *A New Theory of Parallels*, Dodgson erroneously compared the areas of open figures, such as a strip of infinite length and finite width, a sector formed by two infinitely long straight lines intersecting at a point in the plane (an angle), and the entire plane, claiming that the area of a strip of infinite length and finite width is smaller than the area of the sector formed by two infinitely long straight lines intersecting in a non-zero angle, which in turn is smaller than the area of the entire plane. This is incorrect, because open figures have no area: they are infinite in extent and so are not 'measurable'. The fallacies in his reasoning result from treating the measures of non-finite geometrical objects as though they were finite, but it was not until the French mathematician Henri Lebesgue and others developed measure theory early in the 20th century, a theory requiring the non-intuitive conceptualization and definition of a countably additive measure, that the measurement problem could be correctly addressed.[8]

The criticisms of Hugh MacColl

In the 1880s and 1890s Hugh MacColl acted as the main reviewer of mathematical books for the London literary and scientific journal *The Athenaeum*. As such he reviewed anonymously most of Dodgson's mathematical books during that period, including the first three editions of *A New Theory of Parallels*. Dodgson continually replied to MacColl's criticisms in later editions of his works.[9]

We should keep in mind the time period when MacColl was reviewing Dodgson's work on geometry. The non-Euclidean geometry of Nikolai Lobachevskii had begun to be widely known soon after 1863 when Jules Hoüel published his French translation of it. Beginning in 1882 with Moritz Pasch's *Vorlesungen über Neuere Geometrie*, and continuing in 1899 with David Hilbert's *Grundlagen der Geometrie*, a transition began to develop from the older view of an axiom as a generally accepted principle that is a self-evident truth to the modern view that axioms are chosen arbitrarily and deal with indefinable concepts about arbitrarily chosen objects of thought.

Hugh MacColl

MacColl objected to Dodgson's hexagon axiom on several grounds, and continued to criticize it even after Dodgson had modified it in response to his criticisms. In his review of the first edition of *A New Theory of Parallels*, MacColl questioned Dodgson's implied assumption of the *possibility* of the inscribed equilateral hexagon, which he asserted that Euclid did not demonstrate until Book IV, Proposition 15, of the *Elements*.[10]

A year later MacColl clarified his objection to the axiom in his review of the second edition of *A New Theory of Parallels*, after Dodgson had responded specifically (but not by name) to his objection in the preface to his new edition. In this preface, in response to another reviewer, Dodgson had added at the bottom of the page containing his offending axiom:[11]

it is assumed to be *theoretically* possible to inscribe an equilateral Hexagon in a Circle.

Answering MacColl, Dodgson then questioned whether the possibility *needs* to be demonstrated. But MacColl was still not satisfied with Dodgson's statements in the preface to this second edition, responding that Dodgson's assumption conflicts with Euclidean practice and requesting him to keep within Euclid's restrictions.

The third edition of *A New Theory of Parallels* was published in 1890 and was essentially an expanded and revised edition of the book. Dodgson modified his problematic axiom, replacing the hexagon by a square and responding quite sarcastically in the preface to MacColl's criticisms. But MacColl remained dissatisfied, and his much longer review of the third edition criticized Dodgson's use of propositions, as well as postulates, to construct the square. Alluding to Dodgson's sarcastic comments about his criticisms, MacColl went on to say:[12]

Postulates and axioms are *permissions* to do or to assume certain things, and are therefore, the opposite of restrictions. The restrictions of Euclid arise from the paucity of his postulates and axioms, a paucity which [he thinks] regrettable.

Unlike MacColl, Dodgson believed that a psychological element was involved in accepting an axiom, as he explained in the preface to the first edition of his book:[13]

I shall be told, no doubt, that this is too *bizarre* and unprecedented an Axiom – that it is an appeal to the *eye* and not to the reason. That it is somewhat *bizarre* I am willing to admit – and am by no means sure that this is not rather a *merit* than a defect. But, as to its being an appeal to the *eye*, what is 'two straight Lines cannot enclose a space' but an appeal to the eye? What is 'all right angles are equal' [Postulate 4] but an appeal to the *eye*?

It appears from his later writings on geometry that MacColl was influenced by this exchange with Dodgson on the nature of axioms. In Section II, 'Axioms, Inference, Implication', of an article written at the end of 1909 and published posthumously in *Mind* in 1910, MacColl wrote:[14]

What is an axiom? No clear line of demarcation can be drawn between an axiom and any other general proposition or formula that is known and admitted to be true...A proposition that may appear axiomatic to one person may appear doubtful to another, until he has obtained a satisfactory proof of it; after which he treats it as an axiom in all subsequent researches...To an omniscient mind would not all true propositions be equally axiomatic? Would it not be absurd to speak of such a mind as *inferring* B from A? This, of course, is an extreme case, but such cases are precisely those that most effectively test the validity of a principle.

This passage suggests that MacColl had given up his belief that an axiom must express a fundamental self-evident truth, and that he had accepted the idea that people differ in what they view as axioms (as opposed to propositions). He went on to say:[15]

strict Euclideans consider no proof valid...if it takes anything for granted that is not founded on Euclid's twelve axioms, although as a matter of fact, Euclid himself, in several of his formal proofs, tacitly assumes axioms which are absent from his given list.

These were the *restrictions* (referred to earlier) that MacColl had alluded to in his last review of *A New Theory of Parallels*. Further evidence that he may have been influenced by Dodgson's views is his description of axioms in the same paper, calling them 'fundamental formulae of appeal';[16] this term is difficult to understand, except when cast in the light of Dodgson's statement that his axiom was an appeal to the eye rather than to the mind. These were the *permissions* that MacColl had also alluded to.

In view of Dodgson's (and MacColl's) lack of clear and correct ideas about the nature of infinitesimals, it is remarkable that Dodgson arrived at the role that the Archimedean axiom plays in a geometry. Using Euclid's definitions of ratio (relative magnitude) in Book V, Dodgson reasoned correctly that this definition was meant to exclude the relation of a finite magnitude to an infinitely great or infinitely small one, where both are magnitudes of the same kind – and he saw that Euclid employed this definition in his proof of Book X, Proposition 1. So Dodgson was one of the few mathematicians of his time to understand that the Archimedean axiom excludes infinitesimals, and by implication that hyperbolic geometry (which admits infinitesimals) is necessarily non-Archimedean. As Dodgson wrote:[17]

My conclusion, then, is that, in Book X. Prop. 1, Euclid is limiting his view to the case of two homogeneous Magnitudes *which are such that neither of them is infinitely greater than the other…* that he is contemplating *Finite Magnitudes only.*

Later work

The importance of Dodgson's discovery became apparent in the independent work of Giuseppe Veronese. Just as we can exclude the parallel axiom from the rest of the axioms of Euclidean geometry in order to investigate what other geometries are possible, so we can exclude the Archimedean axiom and investigate non-Archimedean geometries, of which Veronese's 1890 geometry was the first.[18]

In 1868 Eugenio Beltrami proved that the parallel axiom is independent of the rest of the axioms of Euclidean plane geometry, so that including either the Euclidean or the hyperbolic parallel axiom yields a consistent geometry. Even so, the acceptance of hyperbolic geometry was slow in coming, and it was not considered a replacement for Euclidean geometry, particularly in English school and university settings.

With his close colleague John Cook Wilson, Oxford's Wykeham Professor of Logic, Dodgson made many attempts to prove that the Euclidean parallel axiom really is a theorem. For Dodgson, the *truth* of Euclidean geometry rested on the consistency of its entire corpus, including its parallel postulate – and that this truth would be unquestionable if its parallel property could be deduced as a theorem.

In Appendix III of the first edition of *A New Theory of Parallels*, Dodgson faulted Euclid's *definition* of parallel lines as lines that do not meet, no matter how far they are produced, because he claimed that it did not produce a *unique* pair of lines, writing:[19]

given a Line and a Point not on it, a whole 'pencil' of Lines may be drawn, through the Point, and not meeting the given Line…after drawing *one* such Line, that the others make with it angles which are *infinitely small fractions of a right angle.*

In other words, allowing parallelism to apply to *infinite* lines introduces both the necessity of infinitesimals and also the possibility of there being *another* parallel axiom – a hyperbolic one.

Trigonometry

In 1861 Dodgson published his only pamphlet on trigonometry, *The Formulae of Plane Trigonometry, Printed with Symbols (Instead of Words) to Express the 'Goniometrical Ratios'*. He had two objectives for it. First, he wanted to include a comprehensive set of trigonometrical formulas in a collection of formulas in pure mathematics that he planned to publish. More importantly, he desired approval for the alternatives that he proposed for the common symbols used for the trigonometrical ratios of sine, cosine, secant, cosecant, tangent, cotangent, and versed-sine (where the versed-sine of x is $1 - \cos x$). For both objectives he sought input from other mathematicians.

His criteria for creating the new symbols were that they should be easily written, suggestive of their meaning, connected with each other, and different from all other currently used symbols. His criteria for constructing each one included requiring just two strokes of the pen, with one of the two lines being the same for all the symbols, but instead of using two lines, he used one line for each. A semicircle was common to all the symbols.

sine   cosine   secant   cosecant   tangent   cotangent   versed-sine

Dodgson's trigonometrical pamphlet was in three parts:

Part I: Measurement of angles by angular units
Part II: Indication, but not measurement, of angles by their goniometrical ratios
Part III: Trigonometry. Properties of rectilinear figures.

In Part II he distinguished a geometrical angle from the angle of position and from the angle of revolution. In Part III he presented many formulas, including Heron's results for triangles, De Moivre's theorem, the infinite series for *e*, and three series of Gregorie, Machin, and Euler for calculating π. Also included were formulas of sines and of tangents, as well as many others.

Algebra

As we saw in Chapter 3, Dodgson published *An Elementary Treatise on Determinants with Their Application to Simultaneous Linear Equations and Algebraical Geometry* in 1867. In his Preface to the book he described its object as follows:[20]

to present the subject as a continuous chain of argument, separated from all accessories of explanation or illustration, a form which I venture to think better suited for a treatise on exact science than the semi-colloquial semi-logical form often adopted by Mathematical writers. I say 'semi-logical' advisedly, for nothing is more easy than to forget, in an argument thus interwoven with illustrative matter, what has, and what has not, been proved.

About Chapter III he wrote:

A complete analysis of a system of simultaneous Linear Equations has always appeared to me to be a desideratum in Algebra: the subject is only touched on in Baltzer; a more complete attempt will be found in Peacock's Algebra, but I have nowhere seen anything like an exhaustive analysis. This chapter aims at furnishing this…

and as an example of Dodgson's penchant for precision we recall from Chapter 3 the following:

I am aware that the word 'Matrix' is already in use to express the very meaning for which I use the word 'Block'; but surely the former word means rather the mould, or form, into which algebraical quantities may be introduced, than an actual assemblage of such quantities…

Condensation and the Alternating Sign Matrix Conjecture

Dodgson made important mathematical discoveries, several of which were not properly recognized until the second half of the 20th century. Of these his 'condensation method' for evaluating determinants (described in Chapter 3) may have had the greatest influence on subsequent mathematical discoveries.[21]

Although Dodgson's condensation method had been used in some linear algebra texts published early in the 20th century, it emerged from relative obscurity in 1986 when David Robbins and Howard Rumsey studied a variation of Dodgson's condensation algorithm in the form of a recurrence that they had created. Notably, it played a seminal role in their discovery of the alternating sign matrix conjecture.[22]

An *alternating sign matrix* is a matrix in which every entry is 0, 1, or −1, and where the non-zero entries in each row and column alternate, starting and ending with 1, and have sum 1. An example of a 4 × 4 alternating sign matrix (there are 42 in total) is:

$$\begin{pmatrix} 0 & 0 & 1 & 0 \\ 1 & 0 & 0 & 0 \\ 0 & 1 & -1 & 1 \\ 0 & 0 & 1 & 0 \end{pmatrix}.$$

Robbins and Rumsey enumerated the alternating sign matrices of a given size. Uncovering a sequence that was new to them:

1, 2, 7, 42, 429, 7436, 218,348, 10,850,216, 911,835,460,…,

they proved that the number of $n \times n$ alternating sign matrices is

$$\frac{1! \cdot 4! \cdot 7! \cdot \ldots \cdot (3n-2)!}{n!(n+1)! \ldots (2n-1)!}.$$

In his book on determinants Dodgson had provided what he called an 'algebraical proof of the condensation method'. It was not a proof, but merely a demonstration for up to 4 × 4 matrices.[23] As we saw in Chapter 3, Dodgson recognized the problems that would be caused by the occurrence of zeros in the interior of any derived block, but believed that it

will be found in practice that…the whole amount of labour will still be much less than that involved in the old process of computation.

In 1999 David Bressoud and James Propp wrote:[24]

Although the use of division may seem like a liability, it actually provides a useful form of error checking for hand calculations with integer matrices: when the algorithm is performed properly… all the entries of all the intervening matrices are integers, so that when a division fails to come out evenly, one can be sure that a mistake has been made somewhere. The method is also useful for computer calculations, especially since it can be executed in parallel by many processors.

We now state Dodgson's determinantal identity in modern terms. For an $n \times n$ matrix A, we let $A_r(i, j)$ be the $r \times r$ minor consisting of r contiguous rows and columns of A, beginning with row i and column j. Then $A_n(1, 1) = \det A$, while $A_{n-2}(2, 2)$ is the central minor and $A_{n-1}(1, 1)$, $A_{n-1}(2, 2)$, $A_{n-1}(1, 2)$, and $A_{n-1}(2, 1)$ are the north-west, south-east, north-east, and south-west minors. Then *Dodgson's determinantal identity* is

$$A_{n-2}(2, 2) A_n(1, 1) = A_{n-1}(1, 1) A_{n-1}(2, 2) - A_{n-1}(1, 2) A_{n-1}(2, 1).$$

It can be shown that the determinant of A can be expressed as the ratio of the difference of the products of two pairs of minors to the central minor. This ratio is a rational function of all its connected minors of any two consecutive sizes.[25]

Dodgson condensation and combinatorics

In 1997 Doron Zeilberger published a combinatorial proof of Dodgson's determinantal identity, accompanied by a Maple package containing the programs that implement the main mappings. He called his theorem, 'Dodgson's determinant evaluation rule proved by two-timing men and women'.[26]

Zeilberger's proof inspired Richard Brualdi and his student Adam Berliner to extend it to what they called the 'Dodgson/Muir combinatorial identity'; this identity, which Thomas Muir had proved in 1883 under the name of 'the law of extensible minors in determinants', includes Dodgson condensation as a special case. Their theorem asserts that a homogeneous determinantal identity for the minors of a matrix remains valid when all the index sets are enlarged by the same disjoint index set.[27]

The *q-analog* of a result is a generalization that introduces q as a new parameter; the q-analog reduces to the original evaluation when $q \to 1$. In 1996 Zeilberger proved algebraically that a q-analog of an important determinant evaluation concerning the enumeration of plane partitions, first developed by Major Percy MacMahon in 1915–16, is a direct result of Dodgson's determinantal identity. Referring to the Italian mathematician Enrico Bombieri, Zeilberger entitled his paper, 'Reverend Charles to the aid of Major Percy and Fields-Medalist Enrico'.[28]

In 2001 Tewodros Amdeberhan and Zeilberger used Dodgson's determinantal identity to justify fifteen explicit determinant evaluations that had been conjectured and then

Doron Zeilberger

proved using computer-assisted methods; the Maple package *Lewis* accompanying their paper automates a few of these.[29]

Finally, in a paper appearing in 2003, Zeilberger showed that if you are trying to solve a problem that involves evaluating a determinant, and if you can guess the required result correctly, then Dodgson's determinantal identity permits an inductive proof of that evaluation – and even the evaluation itself can be proved by using Dodgson's identity. He also discussed where this paper, together with his three previous ones and his combinatorial proof of Dodgson's identity, were leading. As he wrote, we need methodologies for creating new algorithms to enable computers both to discover and to prove new results.[30]

In another extension of Dodgson condensation, in perhaps his last paper (published in 2005), David Robbins described a non-Archimedean approximate form of the method and stated the following conjecture which he checked in billions of cases:[31]

Suppose a determinant is computed with approximate n-digit floating point Dodgson condensation. If the condensation error for the computation is e, then, after conversion to fixed point, the result will be correct mod p^{n-e}.

Recall that each application of Dodgson's determinantal identity requires a division by a previously computed determinant – that of C, the central minor. While computing a given minor, Robbins defined the *condensation error* of that minor to be e if the maximum of the exponents of all the divisors of det C is e. The matrix entries are n-digit floating point integers – that is, each is a pair (a, e), where a is invertible modulo p^n and e is a non-negative integer; a and e play the roles of the mantissa and exponent in ordinary floating-point arithmetic.

Quasi-determinants

Quasi-determinants are defined differently from determinants. They can be considered as non-commutative analogues of the ratio of a determinant to one of its principal minors. In a landmark paper, published in 1991, I. M. Gelfand and V. S. Retakh constructed general quasi-determinants of matrices over non-commutative rings.[32] Quasi-determinants have become an important topic in non-commutative algebra.

Defining the determinant of a matrix as a polynomial function of its entries does not work for quasi-determinants. Algorithms of Sylvester and Dodgson are feasible recursive methods for computing both determinants and quasi-determinants, and from a foundation point of view both can be expressed in terms of ratios – that is, quasi-determinants are non-commutative analogues of the ratio of the determinant of an $n \times n$ matrix to the determinant of a suitable $(n-1) \times (n-1)$ submatrix.

As Gelfand and Retakh remarked, their non-commutative version of Sylvester's identity applies also to the case of a square submatrix of A composed of some rows and columns of A that are not necessarily consecutive and are not necessarily the same rows and columns of A. This generalization, first reported by them in 1997, is an analogue of a well-known commutative identity... the 'Lewis Carroll identity'. This was the first time that Dodgson's identity had been identified as a special case of Sylvester's identity.[33]

Like Sylvester's identity, Dodgson's identity provides a recursive method for computing both determinants and quasi-determinants, by reducing their computation to that of computing 2×2 determinants or quasi-determinants. As we have indicated, computing quasi-determinants for large matrices is even more tedious than computing determinants.

The identities of both Sylvester and Dodgson can be programmed to run in cubic polynomial time, and that time can be further improved when they are run on parallel processors. Another important aspect of these identities is that both deal with block matrices – that is, with the decomposition of a square matrix A into its submatrices.

Logic

Dodgson began writing explicitly about logic in the 1870s when he began his magnum opus *Symbolic Logic*, with the first part appearing in 1896 (see Chapter 4). Logical arguments using rules of inference are a major component of both geometry and logic, sharing the characteristics of truth and certainty, and in a letter to John Alexander Stewart, Oxford's White Professor of Moral Philosophy, Dodgson wrote of logic:[34]

It is its absolute *certainty* which at present fascinates me.

Dodgson's formulation of formal logic came late in his life, following his publications on Euclid's geometry in the 1860s and 1870s. From the mid-1880s onwards, Dodgson shifted his focus from the truth given by geometrical theorems (true statements) to the validity of logical arguments (the rules that guarantee that only true conclusions can be inferred from true premises), and he pushed the envelope of the standard forms of the prevailing Aristotelian logic of his time.

Although Dodgson worked alone, he was not isolated from the community of logicians. As we saw in Chapter 4, he corresponded with many British logicians, including John Cook Wilson, James Welton, Thomas Fowler, William Ernest Johnson, Herbert William Blunt, Henry Sidgwick, John Alexander Stewart, and Bartholomew Price.[35] In Book XXII, 'Solutions of problems set by other writers', in the edition by W. W. Bartley of *Lewis Carroll's Symbolic Logic*, Dodgson referred to the work of many of the best logicians of his time: George Boole, Augustus De Morgan, William Stanley Jevons, John Neville Keynes, John Venn, and C. S. Peirce's students.[36]

Dodgson was both a popularizer and an educator. Throughout his career he gave private lessons on mathematical topics to small groups of children, and sometimes to their parents and teachers. In 1887 he began teaching logic classes at Oxford's high schools, continuing this activity intermittently up to July 1896. The topics that he chose to teach privately focused on memory aids, number tricks, computational shortcuts, and problems suited to rapid mental calculation, developing this last topic into a book, *Curiosa Mathematica, Part II. Pillow-Problems*, published in 1893 (see Chapter 6). He believed that mental activities and recreations, such as games and puzzles, were enjoyable and conferred a sense of power on those who make the effort to solve them. In an Appendix to the fourth edition of *Symbolic Logic, Part I*, addressed to teachers, he wrote:[37]

Symbolic Logic has one *unique* feature, as compared with games and puzzles, which entitles it, I hold, to rank above them all… [The accomplished *Logician*] may apply his skill to any and every subject of human thought; in every one of them it will help him to get *clear* ideas, to make *orderly* arrangement of his knowledge, and more important than all, to detect and unravel the *fallacies* he will meet with in every subject he may interest himself in.

Why did Dodgson choose to write his logic books under his pseudonym? Bartley suggests a combination of motives. First, Dodgson wanted the material to appeal to a large general audience, particularly to young people, a task that was made easier by using the wide acclaim accorded him as the writer Lewis Carroll.

In his logic books especially, his use of humour set his work apart. An anonymous reviewer of *Symbolic Logic, Part I* in *The Educational Times* wrote:[38]

this very uncommon exposition of elementary logic appears to have tickled the fancy of folk.

The quotations that continue to be cited by modern authors, particularly from his logic books, reinforce this view. However, the reaction of Hugh MacColl, who reviewed the book for *The Athenaeum*, was mixed. On 17 October 1896 he described Carroll's diagrammatic method for solving logical problems as 'elegant', but was critical of his subscript notation and of his interpretation that 'All' propositions imply the existence of their subject.

Bartley and *Symbolic Logic*

Dodgson's place in the development of symbolic logic in England was unknown until 1977, when Bartley published his edition of *Lewis Carroll's Symbolic Logic*. This included a construction from some newly discovered manuscript material of Dodgson's unpublished

William Warren Bartley, III and *Lewis Carroll's Symbolic Logic*

second part of that book, and in consequence Bartley's publication began to change Dodgson's reputation as a logician. In the second edition he included additional galley proof discoveries, solutions to some of Dodgson's more significant problems and puzzles, and a new interpretation by Mark Richards of Dodgson's logical charts.[39]

A comparison of the two parts of *Symbolic Logic* reveals the progress that Dodgson had made toward an *automated* approach to the solution of puzzle problems. Although Dodgson worked with a restricted form of the logic of classes and used rather awkward notation and odd names, he introduced important methods that foreshadowed modern concepts and techniques in automated reasoning.

As we saw in Chapter 4, the barber-shop paradox was Dodgson's first publication in the influential journal *Mind*. It is the transcription of a dispute that placed him in opposition to Oxford's Professor of Logic, John Cook Wilson. John Venn had been one of the first to discuss it in print, in the second edition of his *Symbolic Logic*, and Bertrand Russell used the paradox in his *Principles of Mathematics* to illustrate his principle that a false proposition implies all others.[40]

Bartley's book includes eight versions of the barber-shop paradox, together with extensive commentary and a discussion of Dodgson's other contribution to *Mind*, 'What the Tortoise Said to Achilles', thereby making the issues they involved more accessible to modern logicians. Each of these articles became the focus of considerable debate within the academic community.

Bartley made several claims about the importance of Dodgson's work, noting that his tree method for reaching valid conclusions from a sorites or puzzle problem was the earliest modern use of what is essentially a *truth tree* to reason efficiently in a multi-literal sorites in the logic of classes. It is a mechanical test of validity using a formal *reductio ad absurdum* argument that bears a close resemblance to the trees employed in the method of 'semantic tableaux' that was published in 1955 by the Dutch logician, Evert W. Beth.[41] In a comprehensive article concerning Dodgson's tree method, Irving Anellis wrote:[42]

Perhaps this valuable contribution to proof theory ought to be called the Hintikka–Smullyan tree method, or even the Dodgson–Hintikka–Smullyan tree…

Although the ingenuity of the puzzles and examples that Dodgson created were generally applauded, Bartley's claims about the significance of Dodgson's work were questioned, so that its value in the development of logic was not fully appreciated when the book was first published. But subsequently, other scholars working on Dodgson's logic writings – such as George Englebretsen (from 1975), Edward Wakeling (1978), Anthony Macula (1995), Mark Richards (2000), and Amirouche Moktefi (from 2007) – have made important discoveries.

Dodgson and symbolic logic

Given that Dodgson's logic publications did not influence the development of symbolic logic in England, how can we then describe his contributions to symbolic logic from a 21st-century point of view? Considering first Bartley's edition of *Lewis Carroll's Symbolic Logic*, we can see that in Parts I and II Dodgson developed a *formal logic* – that is, a logic in which the validity of an argument is a matter of form only, and not of interpretation – by setting down intuitively valid formal rules for making inferences. Since a variety of possible steps are available when carrying out an inference, Dodgson used his rules to determine which steps were the most promising, and the set of techniques that he created are capable of mechanizing logical reasoning.

Dodgson's approach led him to invent various methods that lend themselves to mechanical reasoning. In *Symbolic Logic, Part I* these are the method of underscoring, the method of subscripts, and the method of diagrams, whereas in *Part II* they are the methods of barred premises and barred groups and, most importantly, the method of trees. Dodgson was the first person in modern times to apply a mechanical procedure, his tree method, to demonstrate the validity of the conclusion of certain complex problems.

Using a tree to test the validity of an argument consisting of two premises and a conclusion, or equivalently determining whether the two premise sentences and the denial of the conclusion sentence are inconsistent by the method of truth tables with (say) three terms, requires calculating eight sets of truth values in order to determine whether a case exists in which all three terms are true. A complete closed tree establishes the validity of the argument by showing there are no cases in which the three sentences are true, but if any path in a finished tree cannot be closed, then the argument is invalid because an open path represents a set of counter-examples.

When Dodgson used the method of barred premises to verify a tree, he guided the generation of the ordered lists by employing an ordering strategy, now known as *unit preference*, which selects first the propositions with the fewest number of terms. He also

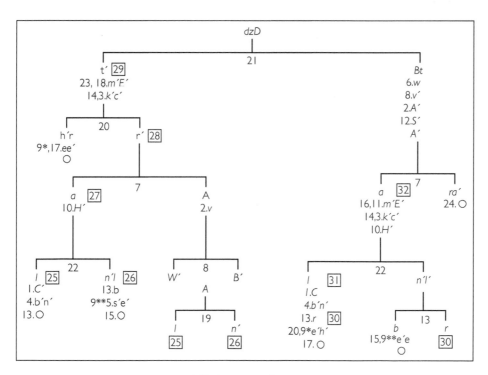

A Dodgson tree diagram

employed a rule to eliminate superfluous premises when verifying a tree. His rule was to ignore such a premise, even if it caused a branching of the tree. But, in the absence of more powerful inference rules and additional strategies, he was unable to approach the solution of these multi-literal problems more efficiently.

Several of the methods that Dodgson used in his *Symbolic Logic* contain kernels of concepts and techniques that were later employed in automatic theorem proving. His only inference rule was *underscoring*, which takes two propositions and selects a term in each of the same subject or predicate with opposite signs to yield another proposition. This is an example of *binary resolution*, the most important of the early proof methods in automated deduction.

What, then, are the shortcomings in Dodgson's work on logic? In his time logicians familiar with *Symbolic Logic, Part I* found serious problems with some of his ideas, particularly that of *existential import*, that 'All' propositions imply the existence of their subject, a concept with which Boole and his followers disagreed; Boole's opposite interpretation is the one that has become accepted. In Dodgson's defence, Bartley observed that he may not have held this idea as a philosophical belief, but that he found it necessary because existential import is implicit in his method of subscripts.

The 'Alice' effect

In an exchange of letters between Venn and Dodgson in 1894, and from the reviews that appeared soon after the publications of *The Game of Logic* and *Symbolic Logic, Part I*, we see that Dodgson's reputation as the author of the *Alice* books cast him primarily as an author of children's books, and this view tended to prevent his logic books from being treated seriously. His own more literary style of writing further contributed to this impression. That this was his reputation was apparent in the reviews of *Symbolic Logic, Part I* in *The Athenaeum* and *The Educational Times*.[43] Certainly most of his contemporaries did not appreciate the importance of the diagrammatic method for solving syllogisms that he first presented in *The Game of Logic*.

As we saw in Chapter 4, Dodgson created the first part of his visual proof system, a diagrammatic system, beginning in 1887 in *The Game of Logic*. His system was capable of detecting fallacies, a subject that greatly interested him, causing him to write:[44]

the Fallacy may be detected by the 'Method of Diagrams', by simply setting them [the propositions] out on a Triliteral Diagram, and observing that they yield no information which can be transferred to the Biliteral Diagram.

With a view to extending his proof method, Dodgson went on to extend his biliteral and triliteral diagrams to more than three sets, eventually creating diagrams for seven and eight sets (classes), and describing the construction of nine-set and ten-set diagrams.

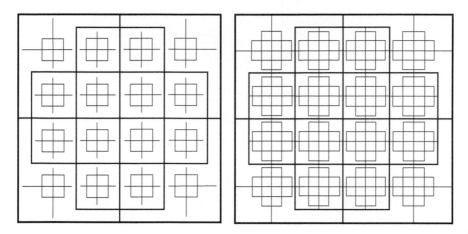

Dodgson's diagrams for seven and eight sets

In 1995 Anthony Macula showed how to construct Dodgson's set diagrams for any number of sets, using a linear iterative process; he called them *Lew k-grams*. More recently, in his 2004 book on Venn diagrams, A. W. F. Edwards wrote:[45]

the Carroll diagram holds the key to the canonical representation of an arbitrary number of sets.

Dodgson's set diagrams are *self-similar*, in the sense that each diagram remains invariant under a change of scale. We see that each k-set diagram has 2^k regions, so that an 8-set diagram has $2^8 = 256$ regions.

Both Venn's diagrams and those of Dodgson are *maximal*, in the sense that no additional logical information (such as inclusive disjunctions) can be represented by them. But because of their self-similarity and algorithmic construction, Dodgson's diagrams are easier to draw for a large number of sets. This regularity makes it simpler to locate (and thereby erase) cells that correspond to classes destroyed by the premises of an argument. Although both types of diagram can represent existential statements, Dodgson's are more versatile than Venn's for easily handling complex problems without compromising the visual clarity of the diagram. Indeed, Dodgson hinted at the superiority of his method when he compared his own solution to a syllogism with one that Venn had supplied.[46]

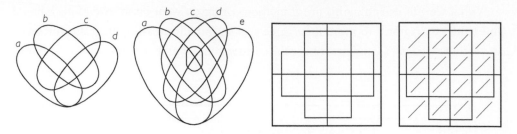

Venn and Carroll diagrams for four and five sets

Existential propositions can be represented in both systems. In Venn's diagrams the use of a small plus sign '+' in a region to indicate that it is non-empty did not appear until 1894, and Dodgson reported on it in *Symbolic Logic, Part I*. But Dodgson may have been the first to use it, since a manuscript worksheet on logical problems, probably dating from 1885, contains a variant of a triliteral diagram that includes a plus sign to represent a non-empty region.[47] However, in his published work Dodgson preferred to use the symbols '1' for a non-empty region and '0' for an empty region.

By his own admission, Hugh MacColl's views on logic were influenced by reading Dodgson's *Symbolic Logic, Part I*.[48] In addition to the clear exposition and the unusual style that characterize Dodgson's books, there seems to be one more essential affinity that supported MacColl's attraction to his work. Their exchanges show that both had a deep interest in the precise use of words, and that both saw no harm in attributing arbitrary meanings to words, as long as the meaning was precise and the attribution was agreed upon.

Unlike Dodgson's earlier publication on the barber-shop paradox, which immediately attracted responses from serious logicians, none was received during his lifetime to 'What the Tortoise Said to Achilles'. As we have seen, this latter work has since been widely discussed among philosophers and is currently considered as a classic text in the philosophy of logic. What is more remarkable is that, in all the articles on this problem that have appeared in journals and books for over a hundred years, there has been no accepted resolution to it.[49]

One of the most important ideas, a consequence of Hilbert's metamathematical framework, is the notion that formalized logical systems can be the subject of mathematical investigation. But it was not until the 1950s that computer programs were used to prove mathematical theorems, using trees as the essential data structures. The focus of these early programs was on proofs of theorems of propositional and predicate logic.[50]

A Disputed Point in Logic

A Concrete Example.

from Verabryne   16/4/94

MS.15

This island consists of a Northern and a Southern Division; but I am not sure where the boundary-line is.

The Northern Division is Brown's estate: the Southern is mine.

Brown is selling his estate to me; but I do not know whether the sale is completed.

The following propositions are true.

I. If this field is Brown's, it must be in the Northern Division (for otherwise it would be part of my estate).

II. If the sale is completed, then, if this field is Brown's, it cannot be in the Northern Division (for otherwise it would be _mine_ by purchase).

Now let "A is true" = "this field is Brown's"

"B is true" = "this field is in the Northern Division"

"C is true" = "the sale is completed"

Then Propositions I, II, are equivalent to (i), (ii), and the question "can C be true?" is equivalent to "is it possible that the sale is completed?"

Here the 2 Propositions, "If A is true B is true" and "If A is true B is not true", both of them contain a logical sequence. Also they are compatible, their combined effect being "A is not true."

Hence, if C is true, A is not true; and, vice versâ, if A is true, C is not true: i.e. A and C cannot be true together.

But there is nothing to prevent C alone being true; i.e. it is possible, consistently with I and II, that the sale _may_ have been completed.

Part of a letter from Dodgson to Cook Wilson on 'A Disputed Point in Logic'

Conclusion

Perhaps the most serious criticism of Dodgson's writings on logic is that they rest entirely on the syllogism and sorites. But, like Boole, Dodgson was a symbolic logician, unlike his Oxford colleague, John Cook Wilson, who opposed the use of mathematics in logic.

Aside from his clever and amusing problems, Dodgson's legacy in logic lies primarily in three areas. First are the methods that he created to mechanize logical reasoning that prefigure methods in modern automated reasoning algorithms. Second, although eclipsed by Venn's diagrammatic system, Dodgson's system of logic diagrams is arguably a better one. Finally, Dodgson's *Mind* article, 'What the Tortoise Said to Achilles', which deals with his paradox of infinity, continues to have significant modern relevance.

Voting theory

Dodgson's publications on mathematical–political subjects offer us a very different picture of the man made famous by the *Alice* books. He was an activist, regularly involved in issues of local governance at Christ Church and in Oxford University, and in problems of national politics at the highest level. To these he applied his extraordinary powers of logical reasoning, augmented by simple mathematical techniques in algebra, probability, and statistics, in order to produce novel solutions to the elements shared by all democratic political processes – voting and elections.

As we saw in Chapter 5, Dodgson was the only 19th-century figure to understand the fundamental issues of the entire spectrum of voting theory. His work remained unknown until more than fifty years after his death, when Duncan Black rediscovered him and made his work available for further investigation.[51] In matters of voting, and in the interests of fairness, Dodgson's greatest concern was the protection of minority opinion. In most of his political writings he expressed this concern by advocating methods that minimized the number of unrepresented voters.

Somewhat surprisingly, game-theoretical ideas surface in most of his political writings, and yet game theory was not systematized into a formal theory until 1928, when John von Neumann published his groundbreaking work on the subject.[52] Even more curiously, Dodgson understood and warned against what he termed *sophisticated*, or tactical, voting. In his 1876 pamphlet he asked whether he *had* to vote for the candidate that he really preferred, or whether he should vote in whatever way would most favour the chances for his preferred candidate to win.[53]

Friendships with Parliamentarians

Dodgson became interested in political matters early and established friendships with several government figures. In a diary entry of 22 September 1857 we first read of his acquaintance with James Garth Marshall MP and his family. Marshall, a Liberal who was very much involved in electoral reform, wrote influential pamphlets on this issue.[54]

Dodgson first met Gathorne Gathorne-Hardy, the Home Secretary, on 8 April 1867, when the latter helped him to enter the House of Commons for the second reading of the Reform Bill which extended the franchise by 88 per cent. On 28 June 1872 he again assisted Dodgson's entry into the House of Commons for the debate on the House of Lords' amendment to the ballot bill that introduced secret voting; prior to this bill, passed in that year, a political party was able to poll within five per cent of its pledged vote because, after an election, printed lists of voters and how they cast their votes were available. On the same day that he met Gathorne-Hardy, Dodgson also met former Christ Church student George Ward Hunt MP and dined with him in the Members' room.

On 3 April 1868 Hunt (by this time Chancellor of the Exchequer) assisted Dodgson in gaining entrance to the House of Commons to hear the debate on the resolution introduced by William Ewart Gladstone (the Leader of the Liberal party) to disestablish the Anglican Irish Church; this resolution would free Roman Catholics from having to pay tithes to support a Church that included only a small part of the Irish population. On 24 April, again with Hunt's help, Dodgson listened to the House debate on this issue. The bill passed in 1869.

Gathorne Gathorne-Hardy and George Ward Hunt

The most important and enduring friendship that Dodgson had with a political figure and his family was with Robert Gascoyne-Cecil, third Marquess of Salisbury and Prime Minister from June 1885 to January 1886, from July 1886 to 1892, and from 1895 to 1902. Their social relationship began in 1870 and lasted throughout Dodgson's life. Salisbury helped Dodgson to gain entry to the House of Lords on 8 July 1872 to hear the debate on amendments to the Ballot bill.

Letters to the *St James's Gazette*

Dodgson frequently gave voice to his opinions in public letters, but it is little known that he engaged in this activity from the beginning of his professional career. Beginning in 1881, Dodgson turned his interest in voting theory away from the local Oxford scene to the wider stage of general elections in Britain, and between 1881 and 1885 he wrote eleven letters to the *St James's Gazette* on aspects of this subject – particularly, proportional representation and redistribution. Four of these letters are entitled 'Proportionate Representation' (see Chapter 5).

In his first letter, dated 15 May 1884, Dodgson criticized the system of voting based on the single transferable vote, advocated by the Proportional Representation Society, which enjoyed wide support among voters and in Parliament. Several distinguished barristers responded, including Arthur Cohen MP, an honours graduate in mathematics from Magdalene College, Cambridge, and one of Britain's distinguished members of the Inner Temple, and William Carr Sidgwick, an acquaintance of Dodgson from their student days, a member of Lincoln's Inn, and a former lecturer in political economy at Oriel College, Oxford. Both favoured the view of John Lubbock, the main architect of proportional representation (see Chapter 5).

Dodgson's fourth letter appeared on 5 June 1884. This example became the core of Section 3 of his pamphlet *The Principles of Parliamentary Representation*, published in October of that year.[55] Dodgson's work on apportionment predates that of Edward V. Huntington who founded the modern theory of apportionment in 1921.[56]

Earlier, in the *St James's Gazette* of 23 March 1882, Dodgson had addressed the problem of cloture (or closure) that had been enacted in February 1882; this gave the House of Commons the right to close debate on an issue and then to vote on it immediately after. Pointedly, and somewhat humorously, Dodgson attacked both the propriety and the logic of cloture, an issue that still resonates today.

Dodgson's principles of fairness and certainty also underlie lawn tennis tournaments, a popular sporting event in Dodgson's time (see Chapter 5), and in Volume II of his book on *The Art of Computer Programming*, Donald Knuth stated that the history of 'minimum comparison selection' can be traced back to Dodgson's essay on the subject.[57] He went on to say that Dodgson's method made more comparisons than necessary, but that it had some interesting aspects for parallel computation and that it appeared to be an excellent plan for a tennis tournament.

Dodgson's legacy in voting theory is substantial: many writers of the 20th and 21st centuries have commented on, and extended, his work on voting.[58]

Probability

Dodgson's work on probability is generally unknown, and there are several reasons for this. First, the state of knowledge of probability in England during his time was not extensive. Second, his publications on the subject appeared in only a few places: *The Educational Times* (from 1885 to 1889), *Pillow-Problems* (from 1893), and rudimentarily in *The Science of Betting* (1866) and *A Tangled Tale* (1885).[59]

Closely related were Dodgson's three letters (1877) to *The Eastbourne Chronicle* on 'Is it well to have our children vaccinated?' In this correspondence he responded to three letters written by the anti-vaccinationist William Hume-Rothery, who had given the number of smallpox patients, observing that 75 per cent of them had been vaccinated. To counter Hume-Rothery's claims Dodgson pressed him to provide accurate statistics of the percentage of Londoners who were vaccinated for the 1870–72 outbreak, and then compared the two percentages. Recently, using a probabilistic argument, Eugene Seneta has established that Dodgson's analysis of the situation was correct.[60]

The science of betting

In 1856 Dodgson discovered the idea of 'pari-mutuel betting', later invented by Pierre Oller in 1865 in France. Pari-mutuel betting is based on the consensus of the subjective probabilities of the betting public who, by placing wagers on the competing horses, determine the odds on each horse: these odds are inversely proportional to the total amount of money bet on each horse. This system guarantees a fixed profit to the operator of the race track, independently of which horses win. As Dodgson noted in his diary for 12 March 1856:[61]

Discovered a principle (probably long known), of making a winning book on any race where *the sum of the chances* (according to market odds) *is not exactly one*. Reduce to a common denominator: put that back into odds, and make your bets in sums proportional to those numbers, *giving all* the odds if the sum *exceeds* one, and *vice versa*.

In a letter dated 19 November 1866 to *The Pall Mall Gazette* he described his system for preventing people from throwing away their money and for guaranteeing winning if they did bet.[62]

Inverse probabilities

There are at least two distinct ways of thinking about probability. The most common is the *empirical view*, where a probability expresses a long-run relative frequency; this approach associates a probability with a sampling distribution estimate. A second way is *inverse probability*, an approach that uses reasoning from observed events to yield the probabilities of the hypotheses that may explain them. A *prior distribution*, from its name, expresses the opinion of the unknown population parameter before the data are available, and a *posterior distribution* expresses the corresponding opinion afterwards. The main tool is *Bayes' Theorem*, a formula for re-evaluating the probability of each hypothesis when one knows that some event has already occurred that produces a posterior distribution of the population. When the sample is small, the prior distribution of the population parameter becomes important, because a different prior distribution produces a different posterior distribution.

In Dodgson's time, those writing on probability issues were Augustus De Morgan in his *Essay on Probabilities* (1838), William Allen Whitworth on *Choice and Chance* (1867), John Venn on *The Logic of Chance* (2nd edn, 1876), and Myron Crofton's essay on 'Probability' in the 9th edition of the *Encyclopedia Britannica* (1885). Dodgson may have come to probability from reading some of these works, and may have first learned of 'strategic voting' from Crofton's article. George Boole also had a strong interest in the subject, primarily as an adjunct to formal logic, and De Morgan, Venn, and Boole all approached probability through logic.

Both De Morgan and Whitworth focused strongly on inverse probabilities, with many of their illustrations involving the drawing of counters from bags. Dodgson's probability problems also tended to be of this kind, with several stages of drawing (see Chapter 6). His approach to choosing a prior distribution for these problems, either a uniform one or a binomial distribution (depending on the wording of the question) was similar to Whitworth's.

On 20 March 1876, Dodgson wrote to Isaac Todhunter in a letter that seems to be the first overt sign of Dodgson's interest in probability, and specifically in inverse probability. In his letter Dodgson asked Todhunter about his solution to an inverse probability problem in the latter's *Algebra* – specifically, the choice of a prior distribution, uniform or binomial, for the number of white balls in a bag containing both black and white balls. Todhunter replied to Dodgson's letter on 24 March.[63] Since Question 38 of

Pillow-Problems was apparently formulated and 'solved' mentally, before Dodgson wrote it out in the same month as the letter had been written, his difficulty with this problem seems closely related to his letter to Todhunter. Question 38 is a difficult one, as we shall see, and Dodgson's solution was incorrect.

Pillow-Problems

As we saw in Chapter 6, Dodgson's *Pillow-Problems* includes thirteen specific Questions relating to probability, and ten of these involve bags and counters (or in more modern terms, *urn models*).[64] There are also two problems in geometrical probability (Questions 45 and 58), and a problem on expectations (Question 10).

We begin with his earliest probability problem, from March 1876, which Dodgson solved incorrectly:

Question 38. There are 3 bags. 'A', 'B', and 'C'. 'A' contains 3 red counters, 'B' 2 red and one white, 'C' one red and 2 white. Two bags are taken at random, and a counter drawn from each: both prove to be red. The counters are replaced, and the experiment is repeated with the same two bags: one proves to be red. What is the chance of the other being red?

Using an approach that facilitated mental calculation, Dodgson first noted that the bag from which the unspecified counter was drawn in the second draw had yielded a red counter in the first draw. He took as his 'causes' the six possible ordered arrangements of A, B, C, the order corresponding to the bag from which the unknown counter was drawn, the bag from which the red counter was drawn twice, and the remaining bag. He rated these causes to be a priori equiprobable, and used Bayes' Theorem to give their posterior probabilities. Since it uses the outcomes from the drawings at the start, this procedure for the prior allocation of probabilities seems incorrect, as was his answer of $^{49}/_{72}$. Instead, a uniform prior distribution $(\frac{1}{3}, \frac{1}{3}, \frac{1}{3})$, taken over the three possible choices of unordered pairs of bags from which the initial choices of counters are made, gives the correct answer of $^{49}/_{95}$.

A different type of probability problem was Question 45:

Question 45. If an infinite number of rods be broken: find the chance that one at least is broken in the middle.

Dodgson's solution to this ill-posed problem from May 1884 is '0.6321207 etc.'. He began by dividing each rod into $n + 1$ parts, where n is odd. Assuming that the n points of division are the only points where a rod can break, that each breaking point is equally likely, and that the number of breaking points is the same as the number of rods – a fallacious assumption – then the probability that one rod does *not* break in the middle is $1 - 1/n$ and the probability that n rods do not break in the middle is $(1 - 1/n)^n$. So the probability that at least one rod breaks in the middle is $1 - (1 - 1/n)^n$, which approaches $1 - 1/e = 0.6321207\ldots$.

In his solution, Dodgson reduced the problem to a finite number of equally likely sample points. The issues in this problem predate some later developments in probability theory, and because of these there was a gap in Dodgson's understanding of the countability and uncountability of collections of numbers. These ideas, which were only beginning to be handled by mathematicians such as Georg Cantor, resulted in a substantial controversy, beginning in March 1885, and highlights Dodgson's understanding of probability in a discrete setting, as we describe next.

The Educational Times

Dodgson contributed three pieces involving probability to *The Educational Times*: *Note on Question 7695* (in 1885), *Something or nothing* (in 1888), and *Question 9588* (in 1889).[65] The motivation for his first piece was Question 7695, posed by J. O'Regan of Limerick:

Question 7695. Two persons play for a stake, each throwing two dice. They throw in turn, A commencing. A wins if he throws a 6, B if he throws a 7: the game ceasing as soon as either event happens. Show that A's chance is to B's as 30 to 31.

The 'reasoning' behind the published solution was as follows. Out of the thirty-six possible outcomes of a toss of two dice, five yield face value 6, giving a probability of $\frac{5}{36}$, and there are six outcomes giving face value 7, with a probability of $\frac{6}{36}$. Thus the odds are

$$\text{' no 7'/'no 6' } = (1 - \tfrac{6}{36}) / (1 - \tfrac{5}{36}) = \tfrac{30}{31}.$$

This answer is correct, but the reasoning is wrong. On 14 March 1885 Dodgson wrote to the editor about the erroneous method of solution, giving a correct argument. If the probability of A's being successful on a single toss is a, and of B's is b, then the probability of A's winning is $a/\{1 - (1-a)(1-b)\}$. The odds of A's winning to B's winning are therefore $a/\{(1-a)b\}$, which with $a = \frac{5}{36}$ and $b = \frac{6}{36}$ gives $\frac{30}{31}$. What is problematic is Dodgson's final statement:

The ratio 30/31, is only *approximative*, the expectations of A and B being just *less* than the fractions 30/61, 31/61. If this were not so, the sum total of their expectations would equal 1; *i.e.* it would be absolutely certain that one or another of them would win — whereas there is clearly a chance, though an indefinitely small one, that the game might go on forever without either winning.

In modern probability the conceivable sample point corresponding to the game going on forever has to be allocated probability 0, since the sum of probabilities of the countable set of sample points on which A or B wins (and so the probability that the game finishes in a finite amount of time) is 1. However, we do not equate the statement that an event has probability 0 with the statement that it cannot occur, as Dodgson does.

Following this, the Revd T. C. Simmons, a competent mathematician, called Dodgson's final statement 'extremely unmathematical'. Dodgson would not concede defeat and both, joined by others, tackled the then uncharted waters of the allocation of probabilities to the sample space [0, 1] to show that *sensible* sets of probability (measure) 0 can readily occur. To this end, in 1886 Simmons sent the following question to *The Educational Times*:

Question 2000: A random point being taken on a given line, what is the chance of it coinciding with a previously assigned point?

Simmons argued correctly that if the point is k, then the probability of taking a point to its left is k and to its right is $1 - k$, so the probability must be 0. Dodgson asserted that if the probability of a specific point is 0, then the probability of any point is 0, yet some point is chosen, and he concluded:

I re-affirm, as absolutely axiomatic, that when an event is *possible*, its chance of happening is *not* zero.

Dodgson would not concede. He countered with another question:

Question 9588: A random point being taken on a given line, find the chance of its dividing the line into two parts [which are] (1) commensurable, (2) incommensurable.

On the basis of Simmons's previous answer of 0, Dodgson argued that Simmons would argue that the probability of selecting a specific rational is 0, and that by adding the probabilities the probability of selecting a rational is 0, as the answer to (1). However, by the same argument, Dodgson said, the probability of an irrational is 0, and hence the probability of any number is 0, so there is a contradiction because *some* point is selected. Dodgson's error was that, in the case of irrationals, he was adding over an uncountable set.

Forty years after this episode, the Soviet mathematician Andrei Kolmogorov formulated three axioms for probability. If we accept these, we would deduce that the probability of selecting an irrational is 1, using the countability of the rationals.

Countability was also at the root of Dodgson's difficulty with the 'game going on forever'. The difficulty reflects his erroneous view of 'infinitesimals': quantities (such as probabilities) that are less than any positive number but are not 0 (so that a corresponding event is not 'impossible'). In Kolmogorov's axiomatization the continuity of countably additive probability measure ensures a probability of 0 for a corresponding limiting event which need not be null (that is, it need not be 'impossible').

John Cook Wilson

As we have seen, John Cook Wilson was an Oxford colleague of Dodgson's. Trained as a classicist and mathematician, he was elected a Fellow of Oriel College in 1873, and in 1889 was appointed Wykeham Professor of Logic at New College. In his geometrical, probabilistic, and logical work, Cook Wilson, though more than twenty-five years younger than Dodgson, was greatly influenced by him.[66]

Eugene Seneta has claimed that Dodgson brought both Myron Crofton's perception of inverse probability and Bayes' Theorem to Cook Wilson, who made an important (but little-known) contribution by pointing out the unifying elements of 'direct' and 'inverse probability', and who may have been one of the first to express what we now understand by the term 'Bayes' Theorem'.[67]

Among Cook Wilson's papers there is a handwritten sheet dated 5 June 1890, signed 'C. L. D.', and entitled *Problem in Chances*. This sheet is the earliest document that has survived of the many letters exchanged between him and Dodgson on topics in geometry and logic. It begins:

There are 2 bags, 'x' *and* 'y', 'x' containing a white and a black ball, 'y' containing a white & 2 black. I draw from one bag, taken at random. If I draw 'w', the thing is over: if not, I am allowed to draw from the other bag. What is my chance of drawing white *somehow*? (to be solved by considering possible events *only*).

This problem involves an application of conditional probabilities. The probability of getting a 'white somehow' can be calculated by using the theorem of total probability, after decomposing the problem into mutually exclusive and exhaustive events. Dodgson's own solution to the problem on the sheet gave the correct final answer of $\frac{2}{3}$. His approach was to consider five 'favourable' but not equally likely events, each giving a 'white somehow' among the eight 'possible' events from repeated drawings, and then to 'estimate' the ratios of probabilities of all eight events to each other.

Cryptology

From the outset of his academic career Dodgson created ciphers, revising the last one in 1888. He based his constructions on the three cipher paradigms of his time: Vigenère, Beaufort, and variant Beaufort, as we describe below.

Dodgson's work on ciphers is not well known. He continued to create anagrams and acrostics well into the 1890s, employing variations inspired by his ciphers, such as varying the location of the 'key' letters in acrostic poems. Together these ciphers, anagrams, acrostics, and poems represent one of the closest associations between his mathematical and his literary interests.[68]

Cryptology is directly tied to word games, and particularly to anagrams and acrostics, both of which hide information. Anagrams are probably closer in spirit to ciphers than acrostics, because the defining ingredient of an anagram is transposition, the changing of the order of a letter which also defines a transposition cipher system. Obscuring word divisions is common to both anagrams and ciphers. Acrostics also have the cryptic quality of concealment, and Dodgson often hid the names of his child friends in them. The use of certain letters to spell out the desired words in an acrostic loosely corresponds to the use of 'key' letters or a 'keyword', which can specify a cipher alphabet or the rearrangement of letters in a transposition.

The basic unit of a code system is a word, whereas that of a cipher is a letter. Generally, cipher systems are of two types: *transposition ciphers*, which change the original letter's order, and *substitution ciphers*, which substitute a letter by another letter or symbol which changes the original letter's form or value. Dodgson created only substitution systems.

The key-vowel cipher

On 23 February 1858, fifteen years after the first electric telegraph had opened in England, Dodgson created his first cipher system. This was part of a simplified (and less secure) form of the Vigenère cipher, a periodic polyalphabetic substitution cipher invented by Blaise de Vigenère in 1585; it was possibly the best-known cipher system before the 20th century.

In the key-vowel cipher the sender uses a keyword to encode a given message, replacing each of its letters by another one. The receiver, who also knows the keyword, then reverses the process to retrieve the original message. But instead of using the twenty-six available alphabets, just five are used, those lines beginning with Y, Z, A, B, C in the following table; these are made to correspond with the *key-vowels*, A, E, I, O, U. A keyword is also required to encode a message using a Vigenère cipher. This word is repeated over the entire message to be enciphered (the *plaintext*). The intersection of the row headed by a letter in the keyword and the column headed by the corresponding letter of the plaintext then produces the *cipherletter*.

	A	B	C	D	E	F	G	H	I	J	K	L	M	N	O	P	Q	R	S	T	U	V	W	X	Y	Z
A	Y	Z	A	B	C	D	E	F	G	H	I	J	K	L	M	N	O	P	Q	R	S	T	U	V	W	X
E	Z	A	B	C	D	E	F	G	H	I	J	K	L	M	N	O	P	Q	R	S	T	U	V	W	X	Y
I	A	B	C	D	E	F	G	H	I	J	K	L	M	N	O	P	Q	R	S	T	U	V	W	X	Y	Z
O	B	C	D	E	F	G	H	I	J	K	L	M	N	O	P	Q	R	S	T	U	V	W	X	Y	Z	A
U	C	D	E	F	G	H	I	J	K	L	M	N	O	P	Q	R	S	T	U	V	W	X	Y	Z	A	B

Dodgson believed that his cipher system was simple enough to memorize, and that it was unbreakable as long as the keyword was kept secret. But how secure was the keyword? It should be short, because the general strategy was to slide the plaintext alongside the ciphertext (assuming that both were available) to locate possible matches, while simultaneously observing the repetitions of pattern that may reveal the length of the keyword. However, recovering the plaintext from the ciphertext alone was practically impossible in his day.

The matrix cipher

Three days later, on 26 February 1858, Dodgson created what he considered to be a much better cipher – still simple enough to memorize and (he believed) unbreakable, provided that the keyword remained unknown – but also one that he claimed would not permit the discovery of the keyword, even if the plaintext were available. To do this he arranged the letters of the alphabet in the form of a 5 × 5 matrix with I also representing J and U representing V:

	0	1	2	3	4
0	A	F	L	Q	W
1	B	G	M	R	X
2	C	H	N	S	Y
3	D	I	O	T	Z
4	E	K	P	U	*

His diaries explain the method of encoding:

Suppose the key-word to be g r o u n d and the first word of the message s e n d.

Measuring from G to S we find it to be "2nd column 1st line," and write 21, in re-translating we begin at G, and go 2 columns to the right, and 1 line further down, and this gives us S again.

Measuring from R to E gives 23, from O to N 04, from U to D 24, i.e. for "send" we write "21, 23, 04, 23."

Using modulo 5 arithmetic we can summarize this as:

if a, b, c, d are any positive integers (row and column numbers), then
$(a, b) + (c, d) = (a + c, b + d) \pmod 5$;

for example, $(2, 3) + (1, 1) = (3, 4) \pmod 5$, or (in letters) $S + G = Z$.

To protect the cipher further he inserted a pair of parenthesized numbers at the beginning – the first a random number, and the second number denoting the *current* position of the keyletter in the encipherment process. The weakness of any cipher system that uses a keyword is the periodicity of that word, and by addressing this vulnerability he guaranteed that his cipher would be practically unbreakable. He also inserted 'nulls' at the beginning and at the end of the ciphertext to obscure word divisions, by following the parenthesized pairs with a letter whose encipherment gave the number of these nulls. Finally, he claimed that by purposely misspelling words (by omitting or adding letters), it would be impossible to recover the keyword from the plaintext.

This was the first cipher to use a non-standard arithmetic. Moreover, it incorporated encipherment instructions within the ciphertext itself – that is, data and instructions – foreshadowing the notion of a stored program that would come almost a century later. Dodgson did not publish either of these first two ciphers: they appear only as diary entries. He made no use of them, and did not take up the subject again for another ten years.

The alphabet-cipher

In 1868 Dodgson reinvented the complete simplified form of the Vigenère cipher, calling it the *alphabet-cipher*. He made similar claims for it as he had done for the ciphers that he had created ten years earlier: the ciphertext cannot be deciphered unless the keyword is known, even if the alphabet table is available – and if not, then it can be reconstructed, making this also an easily memorized system. However, Dodgson made no attempt to protect the cipher, not even obscuring word divisions. He used the cipher primarily in his letters to child friends.

THE ALPHABET-CIPHER.

	A	B	C	D	E	F	G	H	I	J	K	L	M	N	O	P	Q	R	S	T	U	V	W	X	Y	Z	
A	a	b	c	d	e	f	g	h	i	j	k	l	m	n	o	p	q	r	s	t	u	v	w	x	y	z	A
B	b	c	d	e	f	g	h	i	j	k	l	m	n	o	p	q	r	s	t	u	v	w	x	y	z	a	B
C	c	d	e	f	g	h	i	j	k	l	m	n	o	p	q	r	s	t	u	v	w	x	y	z	a	b	C
D	d	e	f	g	h	i	j	k	l	m	n	o	p	q	r	s	t	u	v	w	x	y	z	a	b	c	D
E	e	f	g	h	i	j	k	l	m	n	o	p	q	r	s	t	u	v	w	x	y	z	a	b	c	d	E
F	f	g	h	i	j	k	l	m	n	o	p	q	r	s	t	u	v	w	x	y	z	a	b	c	d	e	F
G	g	h	i	j	k	l	m	n	o	p	q	r	s	t	u	v	w	x	y	z	a	b	c	d	e	f	G
H	h	i	j	k	l	m	n	o	p	q	r	s	t	u	v	w	x	y	z	a	b	c	d	e	f	g	H
I	i	j	k	l	m	n	o	p	q	r	s	t	u	v	w	x	y	z	a	b	c	d	e	f	g	h	I
J	j	k	l	m	n	o	p	q	r	s	t	u	v	w	x	y	z	a	b	c	d	e	f	g	h	i	J
K	k	l	m	n	o	p	q	r	s	t	u	v	w	x	y	z	a	b	c	d	e	f	g	h	i	j	K
L	l	m	n	o	p	q	r	s	t	u	v	w	x	y	z	a	b	c	d	e	f	g	h	i	j	k	L
M	m	n	o	p	q	r	s	t	u	v	w	x	y	z	a	b	c	d	e	f	g	h	i	j	k	l	M
N	n	o	p	q	r	s	t	u	v	w	x	y	z	a	b	c	d	e	f	g	h	i	j	k	l	m	N
O	o	p	q	r	s	t	u	v	w	x	y	z	a	b	c	d	e	f	g	h	i	j	k	l	m	n	O
P	p	q	r	s	t	u	v	w	x	y	z	a	b	c	d	e	f	g	h	i	j	k	l	m	n	o	P
Q	q	r	s	t	u	v	w	x	y	z	a	b	c	d	e	f	g	h	i	j	k	l	m	n	o	p	Q
R	r	s	t	u	v	w	x	y	z	a	b	c	d	e	f	g	h	i	j	k	l	m	n	o	p	q	R
S	s	t	u	v	w	x	y	z	a	b	c	d	e	f	g	h	i	j	k	l	m	n	o	p	q	r	S
T	t	u	v	w	x	y	z	a	b	c	d	e	f	g	h	i	j	k	l	m	n	o	p	q	r	s	T
U	u	v	w	x	y	z	a	b	c	d	e	f	g	h	i	j	k	l	m	n	o	p	q	r	s	t	U
V	v	w	x	y	z	a	b	c	d	e	f	g	h	i	j	k	l	m	n	o	p	q	r	s	t	u	V
W	w	x	y	z	a	b	c	d	e	f	g	h	i	j	k	l	m	n	o	p	q	r	s	t	u	v	W
X	x	y	z	a	b	c	d	e	f	g	h	i	j	k	l	m	n	o	p	q	r	s	t	u	v	w	X
Y	y	z	a	b	c	d	e	f	g	h	i	j	k	l	m	n	o	p	q	r	s	t	u	v	w	x	Y
Z	z	a	b	c	d	e	f	g	h	i	j	k	l	m	n	o	p	q	r	s	t	u	v	w	x	y	Z

An explanation of the method of using the above table for sending Messages will be found on the other side.

The alphabet-cipher
[Richards Collection]

As before, we need a keyword, known only to the sender and the receiver; following Dodgson, let us use the keyword *VIGILANCE*. Then, to send a message such as

MEET ME ON TUESDAY EVENING AT SEVEN

we write out the following table:

V	I	G	I	L	A	N	C	E	V	I	G	I	L	A	N	C	E	V	I	G	I	L	A	N	C	E	V	I
M	E	E	T	M	E	O	N	T	U	E	S	D	A	Y	E	V	E	N	I	N	G	A	T	S	E	V	E	N
H	M	K	B	X	E	B	P	X	P	M	Y	L	L	Y	R	X	I	I	Q	T	O	L	T	F	G	Z	Z	V

– for example, to encode the letter M, write down the letter in row V and column M: this is H. The last line of the table gives the encoded message.

The telegraph-cipher

On 22 April 1868 Dodgson invented another cipher that he would use to communicate with child friends. Printed on a small card his system employed two sliding alphabets, one for the keyletters (the *key-alphabet*) and the other for the plaintext (the *message-alphabet*):[69]

Key – Alphabet

A B C D E F G H I J K L M N O P Q R S T U V W X Y Z

A B C D E F G H I J K L M N O P Q R S T U V W X Y Z

Message – Alphabet

Rules

The correspondents agree on a 'key-word,' which must be kept secret.

To translate a message into cipher, write the key-word over it, letter by letter, repeating it as often as may be necessary. Find the first letter of the key-word in the 'key-alphabet' and the first letter of the message in the 'message-alphabet': bring them into a column by sliding one alphabet under the other, and copy the letter over 'a': this is the first letter of the cipher.

Translate the cipher into English by same rule.

For example, if the key-word is 'WAR' and the message is 'MEET ME AT SIX' then we write:

```
W A R W   A R   W A   R W A
M E E T   M E   A T   S I X
K W N D   O N   W H   Z O D
```

and send the ciphertext 'KWNDONWHZOD'. This can then be re-translated by the same process.

Other than stating that the keyword had to be kept secret, Dodgson made no claims about the security of this cipher, and nor did he add embellishments that would obscure normal word divisions. The result is a *Beaufort enciphered message*, named after the retired British Royal Navy Admiral Sir Francis Beaufort who had reinvented this cipher, probably in 1857.[70] It was first proposed by Giovanni Sestri in 1710.

The last of the main 19th-century polyalphabetic ciphers was derived from the Beaufort cipher and was named the *variant Beaufort cipher*. Its encipherment was given by $C = P - K$, where P is the message and K is the repeated keyword. Comparing this encipherment scheme with the Vigenère *decipherment* process, $P = C - K$, we see that each is the inverse of the other. Using the alphabet table above to simulate a variant Beaufort encipherment, we observe that the matrix cipher is a variant Beaufort system. It is entirely possible that Dodgson knew about Beaufort's work, which was well known at the time, but he did not mention it. On 24 April 1868 Dodgson communicated his invention to George Ward Hunt, then Chancellor of the Exchequer, at a dinner; there is no record of a response.

Although Dodgson believed his ciphers to be unbreakable, this was not the case. In the 19th century few people knew of the existence of the book *Die Geheimschriften und die Dechiffrirkunst* (Secret Writing and the Art of Decipherment), published in Berlin in 1863, in which its author, Friedrich W. Kasiski, gave the first general solution to poly-alphabetic ciphers with repeating keywords.

After 1875 the nature of Dodgson's cipher work changed. He created the first two ciphers (the key-vowel cipher and the matrix cipher) to be suitable for military and diplomatic purposes, and from a practical viewpoint these were unbreakable systems. The alphabet-cipher and telegraph-cipher appeared after the publication of the *Alice* books: they were simpler to use, and for ordinary telegrams and mailed postcards were well constructed and secure by the standards of his time. Twenty-five years after Dodgson had created his first two ciphers, Auguste Kerckhoffs established the basic requirements for secure military ciphers transmitted by telegraph.[71] These were:

the system should be unbreakable in practice
the key should be easily remembered and changeable
the system should be simple, and not involve a long or difficult list of rules.

Dodgson's ciphers of 1858 and 1868 certainly met these standards.

Conclusion

As we have seen, Dodgson's serious mathematical writings range over a wide variety of topics, including geometry, algebra, logic, voting theory, probability, and cryptology. He worked on these topics in his own time, while carrying on his duties as the Lecturer in Mathematics at Christ Church.

The inspiration for much of what he wrote came from his participation in the life of the College as a resident Fellow of Christ Church, and from his contacts with faculty members at other colleges in the University and in other English universities, particularly in Cambridge. He communicated his work within a circle of colleagues and solicited their opinions. Unlike most of his colleagues he did not seek membership in the professional mathematical societies, such as the London Mathematical Society, the Royal Society, or the Oxford Mathematical Society, and nor did he attend their meetings or lecture there or at meetings of other academic groups in Oxford or London; nor did he try to build up his mathematical reputation in other ways.

Why did scholars become interested in Dodgson's serious mathematical work only in the third quarter of the 20th century? There are several reasons, but one of the most important is the role that certain publishers eventually played in making Dodgson's mathematical work available: these included Clarkson N. Potter and Dover Press in New York City, and Kluwer in the Netherlands whose books were distributed in the UK and the USA. The articles in Martin Gardner's popular *Mathematical Games* section of *Scientific American* also included several of Dodgson's mathematical ideas, and were invaluable sources of information for scholars.[72] Another important reason is that only in the 20th century did some of his mathematical ideas find application, in the sense that his work foreshadowed their use.

In order to investigate his mathematical legacy, we have asked a few more questions: What kind of a mathematician was he? Certainly he was not a traditional research mathematician like his colleague Henry Smith or his Cambridge contemporary Arthur Cayley. Rather, he applied mathematical solutions to all sorts of problems that interested him. As a natural logician at a time when logic was not considered to be a part of mathematics, he successfully worked in both fields. For example:

algebra, where he shortened tedious matrix calculations by applying his condensation method;

Parliamentary elections, where he maximized the percentage of the electorate represented by the members of a district for which he formulated bi-proportional representation;

symbolic logic, where he created visual methods to test the validity of arguments.

In summary, Dodgson's mathematical contributions were broadly based and his legacy is a significant one. His influence on mathematics occurred primarily after his death, and aspects of his work influenced important mathematical developments in the 20th century.[73]

CHAPTER 8

Mathematical bibliography

MARK R. RICHARDS

The aim of this bibliography is to provide an overview of the scope of Dodgson's mathematical works in sufficient detail to enable readers to gain a quick grasp of the extent and essential details of his work in each area of study. All of the major works are included, as are all the significant pieces published in journals, and any manuscripts or galley proofs that are relevant to the main subject areas in which he worked. Some minor pieces – such as short manuscripts and descriptions of work in his diaries and letters – are omitted, on the basis that anyone studying the subject at that level will need to consult *all* the published material, as well as unpublished pieces in various libraries and archives. Various fragments of his work on logic and many of his games and puzzles are also omitted, partly for the same reason but in some cases because they might not be regarded as truly mathematical.

So this bibliography has been constructed for the use of general readers and for scholars of Dodgson's mathematical works, rather than for collectors or cataloguers. The latter groups are well served by *The Lewis Carroll Handbook* and *Lewis Carroll and the Press* (see below) and the numerous articles on the minutiae of Dodgson's publications that have appeared in various journals since. While the *Handbook* is recommended for anyone studying Dodgson's life and works, it does not easily facilitate an overall understanding of the scope of his output as a mathematician, and omits a number of items and explanatory details that are useful, even if not necessary, for the advanced study of specific areas of his work.

The Mathematical World of Charles L. Dodgson (Lewis Carroll). Robin Wilson and Amirouche Moktefi.
Oxford University Press (2019). © Oxford University Press 2019.
DOI: 10.1093/oso/9780198817000.001.0001

With the ever-increasing availability of online texts and extensive print-on-demand facilities, almost all of Dodgson's work listed here is readily available, even if for a price. Where possible, this list gives references to a few specific books that are recommended, partly for their accuracy and ease of access (having been carefully edited and having numerous pieces all in one place), and partly because of the additional commentary that accompanies the original work.

The entries in this bibliography list the work's title, the author's name (C. L. Dodgson, Lewis Carroll, etc.), the publisher, and a short commentary where appropriate.

References

Bibliographical references are to the following three works:

Collingwood   *The Life and Letters of Lewis Carroll*
Stuart Dodgson Collingwood
T. Fisher Unwin (1898).

Handbook   *The Lewis Carroll Handbook,*
edited by S. H. Williams, F. Madan, and R. L. Green, and revised by Denis Crutch
Dawson (1979).

Lovett   *Lewis Carroll and the Press*
Charles Lovett
Oak Knoll Press / The British Library (1999).

Other references cited below are:

Abeles I   *The Mathematical Pamphlets of Charles Lutwidge Dodgson and Related Pieces*
compiled, with introductory essays, notes, and annotations by Francine F. Abeles
Lewis Carroll Society of North America (1994).

Abeles 2   *The Political Pamphlets and Letters of Charles Lutwidge Dodgson and Related Pieces*
compiled, with introductory essays, notes, and annotations by Francine F. Abeles
Lewis Carroll Society of North America (2001).

Abeles 3   *The Logic Pamphlets of Charles Lutwidge Dodgson and Related Pieces*
compiled, with introductory essays, notes, and annotations by Francine F. Abeles
Lewis Carroll Society of North America (2010).

Bartley   *Lewis Carroll's Symbolic Logic: Part I, Elementary, 1896, fifth edition, Part II, Advanced, never previously published. Together with letters from Lewis Carroll to eminent nineteenth-century Logicians and to his 'logical sister,' and eight versions of the Barber-Shop Paradox*
William Warren Bartley, III
Harvester Press, 1st edn (1977).

Black	*The Theory of Committees and Elections*
	Duncan Black
	Cambridge University Press (1958).
Fisher	*The Magic of Lewis Carroll*
	edited by John Fisher
	Nelson, London (1973).
Gardner	*The Universe in a Handkerchief*
	Martin Gardner
	Copernicus, Springer-Verlag (1996).
Gattégno	*Lewis Carroll. Fragments of a Looking-Glass. From Alice to Zeno*
	Jean Gattégno
	George Allen and Unwin, London (1977).
McLean et al.	*A Mathematical Approach to Proportional Representation: Duncan Black on Lewis Carroll*
	edited by I. McLean, A. McMillan, and B. L. Monroe
	Kluwer (1996).
Morgan	*The Pamphlets of Lewis Carroll. Games, Puzzles, and Related Pieces*
	compiled, with introductory essays, notes, and annotations by Christopher Morgan
	Lewis Carroll Society of North America (2015).
Wakeling	*Rediscovered Lewis Carroll Puzzles*
	newly compiled and edited by Edward Wakeling
	Dover, New York (1995).

Bibliography

1860

Notes on the First Two Books of Euclid
Designed for Candidates for Responsions
Anonymous
John Henry and James Parker, Oxford (1860).
A collection of definitions relating to Euclid's *Elements*, Books I and II, for candidates for Responsions (the first examination for an Oxford degree).
Reprinted in **Abeles 1**, 35–41; listed in **Handbook**, 18–19.

A Syllabus of Plane Algebraical Geometry,
Systematically arranged, with formal definitions, postulates, and axioms
Charles Lutwidge Dodgson
Printed by James Wright, Oxford; sold by J. H. and J. Parker (1860).
Listed in **Handbook**, 18.

1861

The Formulae of Plane Trigonometry
Printed with symbols (instead of words) to express 'goniometrical ratios'
Charles Lutwidge Dodgson
Printed by James Wright, Oxford; sold by J. H. and J. Parker (1861).
An attempt to introduce a collection of formulas in algebraic geometry and new symbols for the expressions 'sin', 'tan', etc., with the hope of eliciting responses from other mathematicians.
Reprinted in **Abeles 1**, 121–39; listed in **Handbook**, 19.

Notes on the First Part of Algebra
Anonymous
Parker, Oxford (1861).
Listed in **Handbook**, 20.

1862

Circular Letter to Mathematical Colleagues and Proof-sheets for General List of Subjects
Letter signed Charles L. Dodgson, proof sheet anonymous
Proofs printed at Oxford University Press (1862).
These were proof sheets laying out topics in pure mathematics and subsequently published as *General List of Subjects* (see below), with an accompanying letter requesting responses from mathematical colleagues.
Reprinted in **Abeles 1**, 350–71; listed in **Handbook**, 22.

1863

The Enunciations of the Proposals and Corollaries, Together with Questions on the Definitions, Postulates, Axioms, &c. in Euclid, Books I and II
Anonymous
Printed by T. Combe, E. P. Hall, and A. Latham, Oxford (1863).
A list of questions and propositions to be proved, relating to Euclid's *Elements*, Books I and II.
The work was later reused as part of *The Enunciations of Euclid, Books I–VI* (1873).
Reprinted in **Abeles 1**, 42–51; listed in **Handbook**, 23.

General List of Subjects, and Cycle for Working Examples
Anonymous
Printed at the University, Oxford (1863).
A list of the topics taught in pure mathematics with examples to be worked by students.
Reprinted in **Abeles 1**, 352–71; listed in **Handbook**, 22.

1864

A Guide to the Mathematical Student in Reading, Reviewing, and Working Examples
Charles Lutwidge Dodgson
John Henry and James Parker, Oxford (1864).
A two-part pamphlet which tabulated the whole subject-matter of pure mathematics in the form of a syllabus for students, followed by a cycle of examples to be worked.
Reprinted in **Abeles 1**, 372–404; listed in **Handbook**, 25.

1866

Card showing Symbols and Abbreviations for Euclid
Dodgson's diary entry for 25 January 1866 refers to his publication of a card of 'symbols and abbreviations for Euclid'. No copy is known to have survived.
Listed in **Handbook**, 37.

'The science of betting'
Charles L. Dodgson
Pall Mall Gazette (19 November 1866), 3; reprinted with minor omissions in *The Times* (20 November 1866), 9.
An explanation and debunking of a method of betting, the details of which were being offered for sale by Messrs. H. and J. Smith. A letter from Dodgson to *Pall Mall Gazette* (20 November 1866), 3, and reprinted in *The Times* (21 November 1866), 10, corrects an arithmetical mistake in the original letter.
Reprinted in **Abeles 2**, 102–4; listed in **Handbook**, 41, and **Lovett**, 32–3.

'Condensation of determinants, being a new and brief method for computing their arithmetical values'
Charles L. Dodgson
Originally published in *Proceedings of the Royal Society* 15, no. 84 (1866), 150–5, and subsequently reprinted as a pamphlet.
Dodgson's explanation of his 'condensation' method for computing the determinant of a matrix.
Reprinted in **Abeles 1**, 170–80; listed in **Handbook**, 39, and **Lovett**, 33–4.

1867

An Elementary Treatise on determinants with their Application to Simultaneous Linear Equations and Algebraic Geometry
Charles L. Dodgson
Macmillan (1867).
Dodgson's major treatise on determinants, partly derived from R. Baltzer's treatise and incorporating some material from standard works on algebraical geometry, but largely original in its approach and including numerous propositions of Dodgson's own devising.
Listed in **Handbook**, 43.

1868

The Alphabet Cipher and *The Telegraph Cipher*
Anonymous
Published and privately distributed 1868.
Single sheets of card with tables and explanations of two methods of encoding a message using letter substitution. The telegraph cipher is more sophisticated than the alphabet cipher. Dodgson had previously devised these ciphers and described them in his diaries (see **Abeles 1**, 325–40). Reprinted in **Abeles 1**, 341–5; listed in **Handbook**, 46–7.

The Fifth Book of Euclid Treated Algebraically, so Far as It Relates to Commensurable Magnitudes, with Notes
Charles L. Dodgson
James Parker and Co., Oxford and London (1868).
A collection of definitions, axioms, and propositions relating to *Euclid,* Book V, covering the theory of ratios. The work is aimed at candidates for Moderations (the second examination for an Oxford degree).
Reprinted in **Abeles 1**, 52–87; listed in **Handbook**, 47–8.

Algebraical Formulae for the Use of Candidates for Responsions
Anonymous
Printed at the University Press, Oxford (1868).
A list of formulas for use by candidates for the first examination for an Oxford degree.
Reprinted in **Abeles 1**, 181–3; listed in **Handbook**, 48.

Formulae in Algebra
Anonymous
Full details of publication unknown, possibly 1868.
Similar to *Algebraical Formulae for the Use of Candidates for Responsions* (1868), with additional pages presenting more advanced formulas and theorems.
Reprinted in **Abeles 1**, 184–9; listed in **Handbook**, 48.

1870

Algebraical Formulae and Rules for the Use of Candidates for Responsions
Anonymous
Printed at the University Press, Oxford (1870).
An adapted and extended version of *Algebraical Formulae for the Use of Candidates for Responsions* (1868).
Reprinted in **Abeles 1**, 190–4; listed in **Handbook**, 57.

Arithmetical Formulae and Rules for the Use of Candidates for Responsions
Anonymous
Printed at the University Press, Oxford (1870).

A collection of arithmetical definitions, tables of weight conversions, etc.
Reprinted in **Abeles 1**, 235–39; listed in **Handbook**, 57.

Arithmetic Papers
Anonymous
Full details of publication unknown, possibly published between 1870 and 1874.
Three undated examination papers of questions in arithmetic. **Abeles 1** points out that the question types are similar to those in *Examples in Arithmetic* (1874).
Reprinted in **Abeles 1**, 266–72; not listed in **Collingwood** or **Handbook**.

1872

Symbols, &c., to be used in Euclid, Books I and II
Printed at the University Press, Oxford (1872).
Details are known only from **Collingwood** and **Handbook**, 71.

Number of Propositions in Euclid
Printed at the University Press, Oxford (1872).
Details are known only from **Collingwood** and **Handbook**, 71.

1873

Enunciations of Euclid I–VI. Together with Questions on the Definitions, Postulates, Axioms, &c.
Anonymous
E. B. Gardner, E. Pickard Hall, and J. H. Stacy, Printers to the University, Oxford (1873).
Comprises amended and enlarged material based on *The Enunciations of the Proposals and Corollaries* (1863) and material similar to that in *The Fifth Book of Euclid* (1868), with new material for Books III, IV, and VI. The whole forms a summary of the work on Euclid's *Elements* required for Moderations (the second set of examinations for an Oxford degree).
Reprinted in **Abeles 1**, 88–113; listed in **Handbook**, 72.

A Discussion of the Various Methods of Procedure in Conducting Elections
Anonymous
E. B. Gardner, E. Pickard Hall, and J. H. Stacy, Printers to the University, Oxford (1873).
The first of Dodgson's pamphlets on methods of voting. **Collingwood** gives the title incorrectly as *A Discussion of the Various Modes of Procedure in Conducting Elections*.
Reprinted in **Black**, 214–22, and **Abeles 2**, 33–41; listed in **Handbook**, 75.
A single-page supplement on the 'Method of Nomination' was later issued, and is included in **Black** and **Abeles 2**, 42; listed in **Handbook**, 75.

1874

Suggestions As to the Best Method of Taking Votes, Where More Than Two Issues Are to Be Voted On
Anonymous
E. Pickard Hall and J. H. Stacy, Printers to the University, Oxford (1874).
The second of Dodgson's pamphlets on voting procedures.
Reprinted in **Black**, 222–4, and **Abeles 2**, 43–5; listed in **Handbook**, 80.

Examples in Arithmetic
Collected by Charles L. Dodgson
E. Pickard Hall and J. H. Stacy, Printers to the University, Oxford (1874).
A collection of arithmetical problems on a range of topics without solutions.
Reprinted in **Abeles 1**, 240–65; listed in **Handbook**, 80–1.

Preliminary Algebra, and Euclid Book V
Printed at the University Press, Oxford (1874).
Details are known only from the bibliography in **Collingwood**. The **Handbook** suggests that this might refer to *Euclid, Book V. Proved Algebraically* (1874).
Listed in **Handbook**, 81.

Euclid, Book V. Proved Algebraically So Far as it Relates to Commensurable Magnitudes
Charles L. Dodgson
James Parker and Co, Oxford (1874).
The **Handbook** identifies this as distinct from the earlier pamphlet, *The Fifth Book of Euclid Treated Algebraically* (1868), in both the content, treatment of the subject, and the order.
Listed in **Handbook**, 81–2.

An Inconceivable Conversation between S and D
Anonymous manuscript, dated 1874.
An extension of Zeno's paradox of Achilles and the Tortoise. **Abeles 3** points out that this is not an early version of 'What the Tortoise Said to Achilles', as suggested by **Handbook**.
First published in **Gattégno**, 304–5, and reprinted in **Abeles 3**, 186–7; listed in **Handbook**, 82.

1875

Euclid, Books I and II
Edited by Charles L. Dodgson
E. Pickard Hall and J. H. Stacy, Printers to the University, Oxford (1875).
Dodgson's first attempt to produce a new edition of Euclid's *Elements*, Books I and II, with only the minimal changes required to make them suitable for use by students. This edition was produced for private circulation, possibly so that he could elicit the views of his colleagues before producing the final edition, which was published in 1882.
Listed in **Handbook**, 86–7.

1876

A Method of Taking Votes on More than Two Issues
Anonymous
Oxford (1876).
Dodgson's third pamphlet on voting procedures, and according to **Black** his most significant. It was produced for private circulation to elicit the opinions of his colleagues.
Reprinted in **Black**, 224–34, and **Abeles 2**, 46–58.
A 'cyclostyled' sheet, dated 1877, requesting feedback from readers, was produced to accompany the pamphlet.
Reprinted in **Black**, 234, and **Abeles 2**, 59–60; listed in **Handbook**, 88–9, 98.

1877

Algebra
Anonymous
Privately produced 'cyclostyled' sheet 1877.
Twelve questions in algebra. The **Handbook** states that the questions were intended for schoolchildren, but **Abeles 1** points out that the problems are not simple and correspond to section 'A' of *A Guide to the Mathematical Student* (1864).
Reprinted in **Abeles 1**, 195–6; listed in **Handbook**, 97.

Formulae (Group C)
Anonymous
Privately produced 'cyclostyled' sheet, possibly in 1877.
This sheet contains a collection of formulas for ratios, trigonometry, and logarithms. **Abeles 1** points out that the formulas correspond to topics in sections 'G' and 'L' of *A Guide to the Mathematical Student* (1864).
Reprinted in **Abeles 1**, 140–3; not listed in either **Collingwood** or **Handbook**.

1878

Formulae
Anonymous
Privately produced 'cyclostyled' sheet, 1878.
This sheet contains a collection of formulas (series and decimal values) for π, e, and trigonometrical functions. **Abeles 1** points out that the formulas correspond to topics in section 'L' of *A Guide to the Mathematical Student* (1864).
Reprinted in **Abeles 1**, 197; not listed in either **Collingwood** or **Handbook**.

1879

'Practical Hints on Teaching long multiplication worked with a single line of figures'
Charles L. Dodgson
Educational Times 32 (1 November 1879), 307–8.
Dodgson presents his new method of working long multiplication, with examples.
Reprinted in **Abeles 1**, 273–6; listed in **Handbook**, 106, and **Lovett**, 50–1.

Euclid and His Modern Rivals
Charles L. Dodgson
Macmillan (1879).
Dodgson's major defence of the use of Euclid's *Elements* in the teaching of geometry, and his criticism of those presenting alternative methods. Although presented in a light-hearted dramatic format, the work is technically detailed.
A second edition of *Euclid and His Modern Rivals* was published in 1885, with a new short preface and numerous changes to the text.

A Supplement to "Euclid and his Modern Rivals" containing a Notice of Henrici's Geometry was also published in 1885. The material from this supplement was included in the second edition. The second edition was later republished by Dover in 1973, with a new introduction by H. S. M. Coxeter.
Listed in **Handbook**, 101–2,132–3.

1880

'The cats and rats again'
Lewis Carroll
The Monthly Packet (February 1880), 204–5.
In this note Dodgson responds to a problem set in *The Monthly Packet* in September 1879 and answered in December 1879. Dodgson found four possible answers to the original problem, making the point that a problem such as this, in double proportion, cannot always be solved by simple arithmetic alone.
Reprinted in **Fisher**, 140–2, **Wakeling**, 34, 70–1, and **Gardner**, 73–4; listed in **Handbook**, 204–5.

'Romantic problems, a tangled tale'
Lewis Carroll
The Monthly Packet (April 1880 to May 1885).
The first appearance in print of one of the ten 'knots' that form *A Tangled Tale*. Each short story contains one or more mathematical puzzles. Dodgson's solutions, analyses of readers' solutions, and supplementary notes appear in subsequent issues of the journal. After 'Knot II' the words 'Romantic problems' were dropped from the title. The full text, with various amendments and changes of sequence, was published as a separate book, *A Tangled Tale* (1885), which ran to four printings (see below).
Listed in **Handbook**, 108–9, and **Lovett**, 56–79.

1882

'Lawn tennis tournaments'
Charles L. Dodgson
St. James's Gazette (12 August 1882), 5–6; reprinted in *St James's Budget* (19 August 1882), 13.
Letter to the editor, highlighting certain problems with conducting tournaments that use a simple knock-out process and proposing an alternative method.
Reprinted in **Abeles 2**, 81–3; listed in **Handbook**, 111, and **Lovett**, 73–4.
A subsequent letter expanding the subject was published on 1 August 1883 (see below).

Simple Facts about Circle-Squaring
An unpublished treatise on circle-squaring for which the following material exists:
Chapter I (*Introductory*) (1882); *Propositions I and II* (undated).
The manuscript of Chapter I was first published in the special edition of the 1932 exhibition catalogue *Lewis Carroll Centenary in London* (Bumpus, 1932).
Reprinted in **Abeles 1**, 144–7.
Proof sheets of Propositions I and II are reproduced in facsimile in **Abeles 1**, 148–9.

Euclid, Books I and II
Edited by Charles L. Dodgson
Macmillan (1882).
This textbook edition of the first two books of Euclid's *Elements* is based on Dodgson's privately distributed book (with the same title) of 1875, with a new introduction, appendices, and numerous changes throughout. A second edition of the work (Oxford, 1883) has the notable change of replacing a number of symbols with words. A further six editions were published between 1885 and 1889.
Listed in **Handbook**, 119–20.

1883

'Lawn tennis tournaments'
Charles L. Dodgson.
St. James's Gazette (1 August 1883), 5–7, and (21 August 1883), 5–6; the latter was reprinted in *St James's Budget* (25 August 1883), 12.
Dodgson returned to this subject that had been addressed in the same journal a year earlier, with a more rigorous explanation of the issue and more detailed proposals regarding the method of conducting tournaments and of scoring them. This led to a series of letters to the editor from other readers and a second letter from Dodgson. The essence of this work was later developed into a full pamphlet, *Lawn Tennis Tournaments: the True Method of Assigning Prizes*, 1883 (see below).
Reprinted in **Abeles 2**, 84–101; listed in **Handbook**, 111, and **Lovett**, 75.

Lawn Tennis Tournaments: the True Method of Assigning Prizes, with a Proof of the Fallacy of the Present Method
Charles L. Dodgson
Macmillan (1883).

Dodgson's proposed method of conducting lawn tennis tournaments so as to avoid the problem of some of the best players being knocked out of the contest prematurely, thereby denying them a chance of a high placing. Opinions vary on the practicality of Dodgson's method, as highlighted in **Abeles 2**, 71. See above for Dodgson's letters on the same subject to the *St. James's Gazette* in 1882 and 1883.

Reprinted in **Abeles 2**, 71–80; listed in **Handbook**, 120.

1884

'Proportionate representation'
Charles L. Dodgson
St James's Gazette (15 May 1884), 5, (19 May 1884), 5–6, (27 May 1884), 5–6, and (5 June 1884), 6.
Dodgson's first letter to the editor highlighted a 'serious defect' in a system of taking votes proposed by 'The Proportionate Representation Society'. This was followed by letters from other readers, prompting three further letters from Dodgson.
Reprinted in **Abeles 2**, 145–54; listed in **Handbook**, 111, and **Lovett**, 76–7.

'Parliamentary elections'
Charles L. Dodgson
St James's Gazette (5 July 1884), 5–6.
Dodgson's proposals for parliamentary representation and processes for making and counting votes.
Reprinted in **Abeles 2**, 156–60; listed in **Handbook**, 111,129, and **Lovett**, 77.

'Redistribution'
Charles L. Dodgson
St James's Gazette (11 October 1884), 3–5.
A further proposal for the election of Parliament, following on from Dodgson's letter to the same journal on 'Parliamentary elections', earlier in the year. A letter to the editor from another reader elicited a short response from Dodgson on 22 October 1884.
Reprinted in **Abeles 2**, 165–75; listed in **Handbook**, 111, and **Lovett**, 77–8.

The Principles of Parliamentary Representation
Charles L. Dodgson
Harrison and Sons, privately distributed, 1884.
A thoroughly argued series of proposals, backed by mathematical examples, for the constitution of parliament and conduct of elections. This pamphlet was rewritten and republished in 1885, with the most technical parts removed to the Appendix; a 'Supplement' and a 'Postscript to the supplement' also appeared in 1885.
Reprinted in **Abeles 2**, 176–208, and with extensive analysis in **McLean *et al.***; listed in **Handbook**, 129–30.

'Divisibility by seven'
C. L. Dodgson
Knowledge (4 July 1884).

An explanation of the method used by Dodgson, taught to him by his father, for ascertaining the divisibility of a number by 7. Dodgson's proof sheet of the letter has survived.

Reprinted in **Abeles 1**, 277.

1885

'Note on Question 7695'

C. L. Dodgson

Educational Times 38 (1 May 1885), 183, and expanded in the supplementary journal, *Mathematical Questions and Solutions from the Educational Times* 43 (January–June 1885), 86–7.

Dodgson's response to the solution to a problem (no. 7695) in probabilities, where the answer was correct, but the method of solution was flawed.

Reprinted in **Abeles 1**, 209–12, and **Lovett**, 78–9.

Dodgson's response led to some private correspondence with the Editor, eventually leading to an article, 'Infinitesimal or zero?' by the Editor in *Mathematical Questions and Solutions from the 'Educational Times'* 44 (July–December 1885), 24–7; here the Editor quotes extensively from Dodgson's correspondence.

Reprinted in **Abeles 1**, 213–18; listed in **Lovett**, 79–80.

In June 1888 Dodgson revisited the subject with a further piece in *The Educational Times* 41. (1 June 1888), 245, entitled 'Something or Nothing'.

Reprinted in **Abeles 1**, 219–20; listed in **Handbook**, 106, and **Lovett**, 87.

A Tangled Tale

Lewis Carroll

Macmillan (1885).

Dodgson's collection of linked short stories was originally published in *The Monthly Packet* from 1880 to 1885 (see above). Each chapter (described as a 'knot') contains one or more embedded problems from various branches of mathematics. Dodgson's solutions and supplementary notes were provided, as were all his comments on (and analyses of) the solutions sent in when the stories were first published in the journal. The sequence of the problems and solutions were changed in a few places, and there were a number of minor textual changes. The first printing of 1000 copies was followed by three separate printings of 1000, each in 1888, with the text remaining constant.

The book was republished by Dover (1958), together with *Pillow-Problems*, although the two books are very different in style and in the level of mathematical ability required to solve the problems, *Pillow-Problems* being more advanced and non-literary. *A Tangled Tale* has been reprinted many times by other publishers (although sometimes omitting the important notes that accompany the solutions), and has been translated into several languages.

Listed in **Handbook**, 136–8.

1886

[Papers on logic]
Published anonymously and privately distributed.

Between 1886 and 1892, Dodgson wrote a number of papers on logic. These were essentially examination papers, mostly for the use of his logic pupils at various schools and colleges. They include:

(A) *First paper on logic* (1886)
Questions on elementary logic propositions and diagrams.
Reprinted in **Abeles 3**, 202–3.

(B) *Second paper on logic*
No known copies.

(C) *Third paper on logic*
No known copies.

(D) *Fourth paper on logic* (1886)
Pairs of premises, each of two terms, from which conclusions are to be drawn where possible.
Reprinted in **Abeles 3**, 204–7.

(E) *Fifth paper on logic* (1887)
Sets of three to twelve premises, each with two to four terms, from which conclusions are to be drawn where possible.
Reprinted in **Abeles 3**, 208–12.

(F) *Sixth paper on logic* (1887)
Sets of five to thirty-nine premises, each with two to four terms, from which conclusions are to be drawn where possible.
Reprinted in **Abeles 3**, 213–17.

(G) *Questions in logic* (possibly 1887)
A set of elementary questions on the propositions, and their representations by Dodgson's diagrams. The paper was reprinted with minor alterations.
Reprinted in **Abeles 3**, 233–45.

(H) *Seventh paper on logic*
No known copies.

(I) *Eighth paper on logic* (1892)
Seven problems of varying types (including 'sequences'), requiring more advanced techniques than were needed for the previous papers. In a variant version of this paper the last two problems were amended.
Reprinted in **Abeles 3**, 218–22.

(J) *Ninth paper on logic* (1892)

Five problems, all but one involving multi-term premises. The last includes forty-four premises from which a conclusion is to be drawn.

Reprinted in **Abeles 3**, 223–7.

(K) *Eighth and ninth papers on logic. Notes* (1892)

Explanatory notes on how the problems in the eighth and ninth papers should be interpreted.

Reprinted in **Abeles 3**, 228–31; listed in **Handbook**, 141, 151.

The Game of Logic
Lewis Carroll
Macmillan (1886).

Dodgson's booklet was aimed at popularizing the study of syllogisms in logic by the use of diagrams and counters. In at least two editions separate printed boards and counters were produced for use with the book. The first printing was withdrawn due to the poor quality of production, and a new edition was issued in 1887.

Republished by Dover in 1958, together with *Symbolic Logic, Part I*, and subsequently by other publishers and now readily available in several languages.

Listed in **Handbook**, 142–3, 149–50.

1887

[Seven logic diagrams]
Anonymous
Undated, but Dodgson's annotations suggest 1887.

Seven diagrams, possibly intended for *Symbolic Logic* (Part II or III), which show connections between different forms of logical proposition and between component parts of them.

Reprinted in **Bartley**, 263–9 (with additional commentary in the second edition of 1986), and in **Abeles 3**, 42–5; listed in **Handbook**, 151.

'To find the day of the week for any given date'
Lewis Carroll
Nature (31 March 1887), 517.

Here Dodgson provides a method for mentally calculating the day of the week for any given date, together with two examples.

Reprinted in **Fisher**, 181–2, **Gardner**, 25–6, and **Abeles 1**, 280–2; listed in **Handbook**, 150.

1888

Curiosa Mathematica, Part I. A New Theory of Parallels
Charles L. Dodgson
Macmillan (1888).

This was Dodgson's attempt to resolve the controversy surrounding Euclid's 12th axiom (or fifth postulate), by replacing it with a simpler axiom of his own devising and developing new proofs

for Euclid's propositions which depended on it. Second (1889), third (1890), and fourth (1895) editions were subsequently issued, each with amendments and corrections. The second edition included a new preface that addressed comments and criticisms made (sometimes publicly) of the first edition. The third edition represents a more significant change in the theory and, in Dodgson's words, is 'greatly simplified and improved'; it includes a new preface that further addresses comments and criticisms received. The fourth edition retains the preface from the third edition and the introduction from the first.
Listed in **Handbook**, 158–9.

'Question 9588'
Charles L. Dodgson
Educational Times 41 (1 July 1888), 280; reprinted in *Mathematical Questions and Solutions from the 'Educational Times'* 50 (July–December 1888), 34–5.
A question in probability set by Dodgson, with his comments on published solutions.
Reprinted in **Abeles 1**, 221–3; listed in **Handbook**, 106, and **Lovett**, 87.

'Question 9636'
Charles L. Dodgson
Educational Times 41 (1 July 1888), 280; reprinted with a solution in *Mathematical Questions and Solutions from the 'Educational Times'* 61 (January–June 1894), 86.
A question on arithmetic computation by Dodgson. Dodgson referred to the problem and solution in his *Curiosa Mathematica, Part II, Pillow-Problems* (1893).
Reprinted in **Abeles 1**, 283–4; listed in **Handbook**, 106, and **Lovett**, 88.

1889

'Question 9995'
Charles L. Dodgson
Educational Times 42 (1 February 1889), 83; reprinted with a solution in *Mathematical Questions and Solutions from the 'Educational Times'* 51 (January–June 1889), 98.
A question in algebra by Dodgson.
Reprinted in **Abeles 1**, 198; listed in **Handbook**, 106, and **Lovett**, 88–9.

Arithmetical Croquet
Anonymous manuscript
Dodgson's number game, to be played in the head or on paper. Originally conceived in 1872, it was later written out for a proposed book of original games and puzzles.
Reprinted in **Fisher**, 103–4, **Gardner**, 39, 42, and **Morgan**, 39–40; listed in **Handbook**, 159.

1892

A Challenge to Logicians
C. L. Dodgson

Printed by Parker (Oxford), privately published.

A single problem on sequences, with eleven premises and a conclusion to be proved.

Reprinted in **Abeles 3**, 109–10; listed in **Handbook**, 179.

'Question 11530'
Rev C. L. Dodgson
Educational Times 45 (1 May 1892), 274; reprinted with a solution in *Mathematical Questions and Solutions from the 'Educational Times'* 59 (January–June 1893), 71–3.
A question in trigonometry by Dodgson.
Reprinted in **Abeles 1**, 150–4; listed in **Handbook**, 106, and **Lovett**, 99–100.

'Rule for Finding Easter-Day for Any Date till A.D. 2499'
Anonymous
Unpublished, corrected pages, possibly between 1892 and 1897.
Dodgson's rule for finding the date of Easter is possibly based on the work of Gauss and Rouse Ball (see **Abeles 1**, 302). The paper includes aids to memory for the method and practice examples.
Reprinted in **Abeles 1**, 302–11.

1893

Curiosa Mathematica, Part II. Pillow-Problems
Charles L. Dodgson
Macmillan (1893).
A collection of 72 questions in various branches of mathematics, with answers and solutions, designed to be solved in the head. In spite of the light-hearted title, most of the questions are challenging. A second edition followed within two months of the first, with numerous amendments, to the point where Dodgson was concerned to ensure that the first edition was withdrawn from sale, even before the second edition was published. Third and fourth editions followed in 1894.
The book was republished by Dover in 1958, together with *A Tangled Tale*, although the two books are very different in style: *A Tangled Tale* is more literary, while solutions to the *Pillow-Problems* require a higher level of mathematical ability.
Listed in **Handbook**, 181–2.

1894

[Papers in symbolic logic]
Anonymous
Published and privately distributed.
In 1894 Dodgson produced further papers of logic questions for use with teaching.

(A) *Symbolic Logic. Questions. I*
Eight elementary questions, including general questions on logic, terms, and the use of diagrams.
Reprinted in **Abeles 3**, 248.

(B) *Symbolic Logic. Questions. II*
Seven further questions, slightly more advanced than the previous set.
Reprinted in **Abeles 3**, 249–50.

(C) *Symbolic Logic. Specimen-Syllogisms. Premisses*
Twenty questions, of which fourteen are simple syllogisms and eight have three or more multi-term premisses. A second version of this paper was produced in which one problem was removed and another was added.
Reprinted in **Abeles 3**, 249–55.

(D) *Symbolic Logic. Specimen-Syllogisms. Solutions*
Solutions to the specimen problems in the revised version of the previous paper.
Reprinted in **Abeles 3**, 256–7; listed in **Handbook**, 187–8.

A Disputed Point in Logic
Although not strictly a paradox, this piece is sometimes referred to as the 'barber-shop paradox' and represents one of Dodgson's most important contributions to the study of logic. Dodgson began his analysis of this specific problem in hypotheticals around 1894, and published various versions of it, including the following. See also 'Question 14122' below.

(A) 'A disputed point in logic' (April 1894)
Anonymous,
Privately distributed.
Reprinted in **Bartley**, 451–2, and **Abeles 3**, 112–13.

(B) 'Disputed point: concrete example' (11 and 16 April 1894)
Two anonymous manuscripts
Reprinted in **Bartley**, 453–4.

(C) 'A disputed point in logic' (May 1894)
Anonymous
Revised version of the earlier paper, privately distributed.
Reprinted in **Bartley**, 454–5, and **Abeles 3**, 114–15.

(D) 'A theorem in logic' (June 1894)
Anonymous
Privately distributed.
Reprinted in **Bartley**, 455–6, and **Abeles 3**, 116–17.

(E) 'A logical paradox'
Lewis Carroll
Published in *Mind* (new series) 3, no. 11 (July 1894), 436–8; also issued as an offprint.
Reprinted in **Bartley**, 456–60, and **Abeles 3**, 118–22; listed in **Lovett**, 102.

(F) 'A logical puzzle' (September 1894)
Anonymous
Privately distributed.
Reprinted in **Bartley**, 460–5, and **Abeles 3**, 123–8.

(G) *A Logical Paradox* (undated)
Galley proofs intended for Part II of *Symbolic Logic*, but never published.
Reprinted in **Bartley**, 428–31, 442; listed in **Handbook**, 88–90.

1895

'Question 12650'
Charles L. Dodgson
Educational Times 47 (1 February 1895), 87; reprinted with solution in *Mathematical Questions and Solutions from the 'Educational Times'* 63 (January–June 1895), 92–3.
A number puzzle by Dodgson, arising out of his interest in divisibility.
Reprinted in **Abeles 1**, 285–7; listed in **Handbook**, 106, and **Lovett**, 102.

'What the Tortoise Said to Achilles'
Lewis Carroll
Mind (new series), 4, no. 14 (April 1895), 278–80; also issued as an offprint, which incorrectly dates the original article as December 1894.
This was Dodgson's second contribution to *Mind*, and is regarded by many as his most important contribution to the study of logic.
Reprinted in **Bartley**, 431–4, and **Abeles 3**, 188–91; listed in **Handbook**, 188, 190, and **Lovett**, 103.

Logical Nomenclature. Desiderata
Anonymous
Privately distributed, 1895.
A list of questions with proposed answers, produced to elicit the views of colleagues on possible terminology for use in logic. Two versions of this paper are known.
Reprinted in **Abeles 3**, 49–52; listed in **Handbook**, 191–2.

1896

Symbolic Logic. Part I. Elementary
Lewis Carroll
Macmillan.
The first and only published part of Dodgson's major three-part treatise on symbolic logic. The second edition was published a few months after the first, with a number of changes, including corrections and changes to definitions. The Appendix (Section 10) includes two new problems as examples of what would be included in Part II. The third edition (described as Second

thousand) was published one month after the second edition, and is essentially the same as before. The fourth edition, issued early in 1897, has a new preface (dated Christmas 1986) and includes number of substantial changes from the third edition; of particular importance is Dodgson's additional emphasis on his definition of propositions 'in A' (propositions of the form 'All x are y'). Changes to the Appendix on the 'existential import of propositions' are also of significance. Five new examples were added to Section 10 of the Appendix, demonstrating the advanced level of problem-making that would have been expected in Part II, had it been published. An advertisement pamphlet for the book and new cards and diagrams (similar to those produced for *The Game of Logic*) were published.

The book was republished by Dover in 1958, together with *The Game of Logic*. A slightly revised version was published as a fifth edition in **Bartley**. Other editions have been published since, and there are a few translations into other languages.

Dodgson died before Parts II and III could be published. Gallery proofs and manuscripts of material prepared for those volumes have been located, and appear in **Bartley**.

Listed in **Handbook**, 194–6.

1897

'Question 13614'
C. L. Dodgson
Educational Times 50 (1 September 1897), 11.
A discussion of a simple formula for finding the number of days in a given month. Dodgson proposed a method which works for all months except February.
Reprinted in **Abeles 1**, 291–2; listed in **Handbook**, 106, and **Lovett**, 103.

'Brief method of dividing a given number by 9 or 11'
Charles L. Dodgson
Nature 56 (14 October 1897), 565–6.
Here Dodgson described his method as though he had recently devised it, although subsequent letters to the journal show that it was not original.
Reprinted in **Abeles 1**, 298–301; listed in **Handbook**, 150, and **Lovett**, 103.
Proof sheets exist from 1897, giving a similar explanation of the method, although these differ in several respects from the article as it appeared in *Nature* (see **Abeles 1**, 293–7).

1898

'Abridged long division'
Charles L. Dodgson
Nature 57 (20 January 1898), 268–71.
A generalization of the methods of division previously published in *Nature* 56 (14 October 1897), and a response to the letters to the editor which the earlier article had generated.
Reprinted in **Abeles 1**, 312–21; listed in **Handbook**, 151, and **Lovett**, 104.

1899

'Question 14122'
The late 'Lewis Carroll'
Educational Times 52 (1 February 1899), 93.
A question in logic by Dodgson, on the 'paradox' discussed in 'A Disputed Point in Logic' (see above). Reprinted with solutions in *Educational Times*, 1 September 1899 and 1 June 1900. Listed in **Handbook**, 106, and **Lovett**, 104.

FURTHER READING, NOTES, AND REFERENCES

Further reading

Abbreviations in square brackets are used in the Detailed notes and references below.

Of the many biographies of Charles Dodgson, one of the best known is:

Morton N. Cohen, *Lewis Carroll: A Biography*, Macmillan (1995). [*Biography*]

There is also much of interest in the biography by his nephew:

Stuart Dodgson Collingwood, *The Life and Letters of Lewis Carroll*, T. Fisher Unwin (1898). [*Collingwood*]

Much useful information about his life can be found in his diaries and letters:

Lewis Carroll's Diaries: The Private Journals of Charles Lutwidge Dodgson, in ten volumes, edited by Edward Wakeling, The Lewis Carroll Society (1993–2007). [*Diaries*]

The Selected Letters of Lewis Carroll, in two volumes, edited by Morton N. Cohen with the assistance of Roger Lancelyn Green, Macmillan (1979). [*Letters*]

There are several collections of his writings. One of the best known is:

The Complete Works of Lewis Carroll, Penguin Books (1988). [*Works*]

Dodgson's main mathematical publications were:

Euclid and his Modern Rivals, Macmillan (1879); reprinted by Dover in 1973. [*Modern Rivals*]

Curiosa Mathematica, Part I. A New Theory of Parallels, Macmillan (1888). [*Parallels*]

An Elementary Treatise on Determinants, with their Application to Simultaneous Linear Equations and Algebraical Geometry, Macmillan (1867). [*Determinants*]

The Game of Logic, Macmillan (1887). [*Game of Logic*]

Symbolic Logic. Part 1: Elementary, 4th edn, Macmillan (1897). [*Symbolic Logic*]

(*Symbolic Logic: Part I* and *The Game of Logic* were reprinted by Dover in 1958.)

There is much information about Dodgson's mathematical activities in:

Robin Wilson, *Lewis Carroll in Numberland: His Fantastical Mathematical Logical Life*, Penguin Books (2009). [*Numberland*]

Lewis Carroll: Man of Science (ed. A. Moktefi, M. Richards, and R. Wilson), *The Carrollian: The Lewis Carroll Journal*, No 30 (October 2017).

Much of Dodgson's Oxford output (both mathematical and of a more general nature) appears in the volumes of *The Pamphlets of Lewis Carroll*, published for the Lewis Carroll Society of North America, and distributed by the University Press of Virginia:

The Oxford Pamphlets, Leaflets, and Circulars of Charles of Charles Lutwidge Dodgson, compiled, with notes and annotations, by Edward Wakeling (1993). [*Oxford Pamphlets*]

The Mathematical Pamphlets of Charles Lutwidge Dodgson and Related Pieces, compiled, with introductory essays, notes, and annotations, by Francine F. Abeles (1994). [*Mathematical Pamphlets*]

The Political Pamphlets and Letters of Charles Lutwidge Dodgson and Related Pieces: A Mathematical Approach; compiled, with introductory essays, notes, and annotations, by Francine F. Abeles (2001). [*Political Pamphlets*]

The Logic Pamphlets of Charles Lutwidge Dodgson and Related Pieces, compiled, with introductory essays, notes, and annotations, by Francine F. Abeles (2010). [*Logic Pamphlets*]

Games, Puzzles, & Related Pieces, compiled, with introductory essays, notes, and annotations, by Christopher Morgan (2015). [*Puzzles Pamphlets*]

Two other books that contain much relevant information are:

Mathematics in Victorian Britain (ed. R. Flood, A. Rice, and R. Wilson), Oxford University Press (2011). [*Victorian Britain*]

Oxford Figures: Eight Centuries of the Mathematical Sciences (ed. J. Fauvel, R. Flood, and R. Wilson), Oxford University Press (2013). [*Oxford Figures*]

Other sources are listed below at the beginning of the appropriate chapters.

Detailed notes and references

CHAPTER 1: A MATHEMATICAL LIFE

The first part of this chapter is adapted from passages in *Numberland*, in which further details can be found. There is also some overlap with Robin Wilson's chapter on Charles Dodgson in *Oxford Figures*.

Notes

1. *Collingwood*, pp. 12–13.
2. *Collingwood*, p. 21.

3. Further examples of problems from Walkingame's book can be found in *Numberland*, p. 30.

4. Dodgson's geometry pages appeared in *The Colophon*, New Graphic Series, No. 2, New York (1939); it is one of the earliest examples of Charles Dodgson's handwriting.

5. *Collingwood*, p. 29.

6. A letter from Charles Dodgson to his sister Elizabeth outlines the revision he needed to do for his Moderations examinations; see *Numberland*, p. 45.

7. *Collingwood*, p. 53.

8. Letter to Mary Dodgson, 23 August 1854, in *Letters* 1, p. 29.

9. *Collingwood*, p. 58.

10. *Diaries*, 22 February 1855, Vol. 1, p. 63.

11. *Diaries*, 24 March 1855, Vol. 1, p. 78.

12. *Diaries*, 30 June 1855, Vol. 1, p. 60.

13. *Diaries*, 26 April 1855, Vol. 1, p. 87.

14. *Collingwood*, p. 59.

15. *Collingwood*, pp. 64–5, and *Diaries*, 31 December 1855, Vol. 1, p. 136.

16. *Diaries*, 12 November 1856, Vol. 2, pp. 113–15.

17. *Diaries*, 26 November 1856, Vol. 2, p. 119.

18. Further information about Dodgson's photography can be found in *Numberland*, pp. 70–5, and in Morton N. Cohen, *Reflections in a Looking Glass: A Centennial Celebration of Lewis Carroll Photographer*, Aperture (1998), and *Lewis Carroll Photographer* (ed. R. Taylor and E. Wakeling), Princeton University Library Albums, Princeton University Press (2002).

19. Further details of this trip can be found in *Numberland*, p. 109.

20. For further information on Dodgson as a deacon and preacher, see *Collingwood*, pp. 74–8.

21. For more information about *The Dynamics of a Parti-cle*, and *The New Method of Evaluation as Applied to π*, see *Numberland*, pp. 87–9 and 124–5, and *Works*, pp. 1011–26.

22. *Collingwood*, p. 218.

23. *Diaries*, 30 November 1881, Vol. 7, p. 381.

24. Duncan Black, *The Theory of Committees and Elections*, Cambridge University Press (1958).

25. See *Political Pamphlets*, p. 31, and Michael Dummett, *Voting Procedures*, Oxford University Press (1986), 5.

26. *Diaries*, 6 September 1855, Vol. 1, p. 129.

27. Introduction to *Symbolic Logic*, pp. xvi–xvii.

28. From *Pillow-Problems*, Introduction to the first edition.

29. *Diaries*, 19 December 1897, Vol. 9, p. 354.

30. See *Diaries*, Vol. 9, p. 357.

31. *Curiosa Mathematica, Part I. A New Theory of Parallels*, 3rd edn (1890), xv–xvi.

32. *Modern Rivals*, p. xi.

33. *Modern Rivals*, p. 143.

34. A. Rice and E. Torrence, '"Shutting up like a telescope": Lewis Carroll's "curious" condensation of method for evaluating determinants', *College Mathematics Journal* 38/2 (2007), 93.

35. The Association for the Improvement of Geometrical Teaching was the first learned association in Britain to be dedicated to the teaching of a specific academic discipline. In 1897 it became the *Mathematical Association*; see M. H. Price, *Mathematics for the Multitude? A History of the Mathematical Association*, Mathematical Association (1994).

36. *Modern Rivals*, p. 182.

37. T. Crilly, *Arthur Cayley: Mathematician Laureate of the Victorian Age,* Johns Hopkins University Press (2006), 323.

38. J. Woolf, *Lewis Carroll in his own Account*, Jabberwock Press (2005), 8–9.

39. See R. B. Sheberman, 'Lewis Carroll and the Society for Psychical Research', *Jabberwocky* 11 (1972), 4–7.

40. See R. Haynes, *The Society for Psychical Research 1882–1982: A History,* MacDonald (1982).

41. J. Stern, *Lewis Carroll Bibliophile*, White Stone (1997), 47.

42. See I. Grattan-Guinness, 'A note on *The Educational Times* and *Mathematical Questions*' and 'Contributing to *The Educational Times*: Letters to W. J. C. Miller', *Historia Mathematica* 19 (1992), 76–8, and 21 (1994), 204–5.

43. See S. E. Despeaux, 'Launching mathematical research without a formal mandate: The role of university-affiliated journals in Britain, 1837–1870', *Historia Mathematica* 34 (2007), 100–2.

44. D. Hudson, *Lewis Carroll*, Constable (1976), 132.

45. *Diaries*, 7 June 1864, Vol. 4, p. 307.

46. See F. F. Abeles, 'Some mathematical letters of Charles L. Dodgson, 1866–1867', *Proc. Canadian Society for History and Philosophy of Mathematics* 10 (1997), 60–7.

47. See A. Moktefi, 'Are other people's books difficult to read? The logic books in Lewis Carroll's private library', *Acta Baltica Historiae et Philosophiae Scientiarum*, 5/1 (2017), 28–49.

CHAPTER 2: GEOMETRY

This chapter is a revised and expanded version of parts of *Numberland*, and of the following articles:

Robin Wilson, 'Charles Dodgson's geometry', *A Bouquet for the Gardener* (ed. M. Burstein), Lewis Carroll Society of North America (2011), 191–204.

Robin Wilson, 'Geometry teaching in England in the 1860s and 1870s: two case studies', *HPM 2004 & ESU4: Proc. ICME10 Satellite Meeting of the HPM Group & The Fourth European Summer University on the History and Epistemology in Mathematics Education* (ed. F. Furinghetti, S. Kaijser, and C. Tzanakis) (2006), 167–73.

Amirouche Moktefi, 'Geometry: the Euclid debate', Chapter 14 of *Victorian Britain*. [*Moktefi*]

The edition of Euclid used is:

Euclid's Elements in the Thomas L. Heath translation (ed. D. Densmore), Green Lion Press (2002). [*Elements*]

and a fuller discussion of the *Elements* is to be found in:

Benno Artmann, *Euclid – The Creation of Mathematics*, Springer (1999).

Dodgson's two main geometrical publications were:

Euclid and his Modern Rivals, Macmillan (1879); reprinted by Dover in 1973. [*Modern Rivals*]

Curiosa Mathematica, Part I. A New Theory of Parallels, Macmillan (1888). [*Parallels*]

Several of his geometry pamphlets are reproduced in *Mathematical Pamphlets*, which also contains a useful and perceptive introduction to Dodgson's geometrical work.

Notes

1. The original version of 'Hiawatha's Photographing' appeared in *The Train* IV (July–December 1857). It exists in several versions and has been reproduced many times.

2. In this chapter we use Sir Thomas L. Heath's translation for quotations from Euclid's *Elements*.

3. Euclid did not originally number his definitions or propositions. We use the numbering proposed by Heath (see note 2); for Victorian readers Robert Simson and other authors used a different numbering.

4. Dodgson's descriptions of a postulate and an axiom appear in his pamphlet *Notes on the First Two Books of Euclid* (1860); see *Mathematical Pamphlets*, p. 36.

5. Examples of ruler-and-compass constructions are methods for bisecting an angle and for constructing a square or regular pentagon; these are given in *Elements*, Book I, Propositions 9 and 46, and Book IV, Proposition 11.

6. See note 4.

7. The letters Q.E.F. (*quod erat faciendum*, what was to be made) were used to indicate the completion of a construction, just as the more familiar abbreviation Q.E.D. (*quod erat demonstrandum*, what was to be shown) was used to indicate the completion of a proof.

8. See *Mathematical Pamphlets*, p. 27.

9. *Diaries*, 16 April 1855, Vol. 1, pp. 82–3.

10. Dodgson's personal copy of Potts's *Euclid* was a later edition, dated 1860.

11. See Montague Burrows, *Pass and Class: An Oxford Guide-Book, through the Courses of Literae Humaniores, Mathematics, Natural Science, and Law and Modern History*, 2nd edn, J. H. and J. Parker (1861), 183–5.

12. See H. MacColl, 'Review of A. E. Layng's *Euclid's Elements of Geometry* (Blackie & Son, 1890)', *The Athenaeum* 3310 (4 April 1891), 443. MacColl's writings on geometry are discussed further in Chapter 7.

13. *Diaries*, 5 March 1855, Vol. 1, p. 70, and 30 November 1881, Vol. 7, p. 381.

14. The quotations appear in Morton N. Cohen (ed.), *Lewis Carroll: Interviews and Recollections*: letters from John H. Pearson and Watkin H. Williams, *The Times* 22 December 1931, p. 6, and 12 January 1932, p. 6.

15. See *Mathematical Pamphlets*, p. 53.

16. Letter from Robert Potts to Dodgson, 20 March 1861.

17. William Whewell, *Of a Liberal Education in General; and with Particular Reference to the Leading Studies of the University of Cambridge*, J. W. Parker (1845), 30.

18. D. Lardner, *The First Six Books of the Elements of Euclid*, 12th edn, Henry G. Bohn (1861), ix.

19. Baden Powell, 'On the present state and future prospects of mathematics at the University of Oxford' (privately printed), Oxford (1832), 40. For more on Baden Powell, see Keith Hannabuss, 'The 19th century', Chapter 10 of *Oxford Figures*.

20. James Pycroft, *Oxford Memories*, Bentley, London (1886), 82.

21. James J. Sylvester, Presidential Address: Mathematics and Physics Section, reprinted as 'A plea for the mathematician', *Nature* 1 (1869–70), 261–2.

22. *Moktefi*, p. 327.

23. [Augustus De Morgan, Review of J. M. Wilson's *Elementary Geometry*], *The Athenaeum* 2125 (18 July 1868), 71–3.

24. B. Russell, 'The teaching of Euclid', *Mathematical Gazette* 2, no 33 (May 1902), 165; see also *Moktefi*, p. 333.

25. The extracts from *Modern Rivals* appearing in this chapter are from the 1885 edition, pages ix–x, 1–2, 11, 209–10, and 225.

26. The Shakespeare quotations appearing in this extract are from *Twelfth Night* (III iv 31) and *The Merchant of Venice* (IV i 335).

27. The discussion of Cooley's book appears in *Modern Rivals*, pp. 60–3.

28. J. M. Wilson, *Elementary Geometry: Part I*, Macmillan (1868), v–xii; see *Moktefi*, p. 335.

29. *Modern Rivals*, p. 42.

30. *Modern Rivals*, pp. 28–30 and 34–6.

31. See *Mathematical Pamphlets*, pp. 17ff.

32. See Stuart Dodgson Collingwood, *The Lewis Carroll Picture Book*, Fisher Unwin (1899), and Dodgson's pamphlet *The Dynamics of a Parti-cle*; see also *Numberland*, pp. 169 and 87–9.

33. *Elements*, Book II, Proposition 14.

34. See Robin Wilson, *Euler's Pioneering Equation*, Oxford University Press (2018), Chapter 3, on π.

35. *Mathematical Pamphlets*, p. 146.

36. *Diaries*, 19 December 1897, Vol. 9, p. 354.

CHAPTER 3: ALGEBRA

Dodgson's book on determinants is:

Charles L. Dodgson, *An Elementary Treatise on Determinants, with their Application to Simultaneous Linear Equations and Algebraical Geometry*, Macmillan (1867). [*Determinants*]

The history of determinants is exhaustively covered in:

Thomas Muir, *The Theory of Determinants in the Historical Order of Development*, 4 vols., Dover (1960). [*Muir*]

For those interested in the history of algebra in general, an excellent resource is:

Victor J. Katz and Karen Hunger Parshall, *Taming the Unknown: A History of Algebra from Antiquity to the Early Twentieth Century*, Princeton University Press (2014): Chapters 11–13 cover 19th-century developments. [*Katz & Parshall*]

Notes

1. See Karen Hunger Parshall, 'Victorian algebra: the freedom to create new mathematical entities', *Victorian Britain*, pp. 339–56.

2. James Joseph Sylvester, 'Additions to the articles in the September number of this Journal, "On a new Class of Theorems," and on Pascal's Theorem', *Philosophical Magazine* 37 (1850), 363–70, on p.369; Arthur Cayley, 'A memoir on the theory of matrices', *Philosophical Transactions of the Royal Society of London* 148 (1858), 17–37.

3. The most comprehensive history of determinants is still *Muir*. For a shorter account, see Eberhard Knobloch, 'Determinants', *Companion Encyclopedia of the History and Philosophy of the Mathematical Sciences*, Vol. 1 (ed. I. Grattan-Guinness), Routledge (1994), 766–74, reprinted by Johns Hopkins University Press (2003).

4. For more on the academic environment at Oxford in this period, see Keith Hannabuss, 'The mid-nineteenth century', *Oxford Figures*, pp. 187–201.

5. Amirouche Moktefi, 'On the social utility of symbolic logic: Lewis Carroll against "The Logicians"', *Studia Metodologiczne* 35 (2015), 133–50, on p. 137.

6. *Determinants*, p. iii.

7. Joseph Gage, 'Undergraduate algebra in nineteenth-century Oxford', *BSHM Bulletin: British Society for the History of Mathematics* 32 (2017), 149–59, on p. 153.

8. At University College London, for example, determinants were introduced to the mathematics curriculum in the late 1860s; see Adrian Rice, 'Mathematics in the metropolis: a survey of Victorian London', *Historia Mathematica* 23 (1996), 376–417, on p. 385.

9. Gage (note 7), p. 153.

10. Gage (note 7), p. 156.

11. *Diaries*, 28 October 1865, Vol. 5, p. 112.

12. *Diaries*, 27 February 1866, Vol. 5, p. 132.

13. *Diaries*, 28 February 1866, Vol. 5, p. 132.

14. Robin Wilson, 'Twinkle, Twinkle, Little Bat!', *The Pembrokian*, No. 35 (July 2011), 16–17; see also Edward Wakeling, 'Lewis Carroll and the Bat', *Antiquarian Book Monthly Review* 9 (7), Issue 99 (1982), 252–9.

15. *Diaries*, 25 March 1866, Vol. 5, p. 133.

16. *Diaries*, 29 March 1866, Vol. 5, p. 133.

17. Letter from Spottiswoode to Dodgson, 2 April 1866, quoted in *Mathematical Pamphlets*, p. 170.

18. *Diaries*, 7 May and 12 May, Vol. 5, pp. 147–9.

19. Charles L. Dodgson, 'Condensation of Determinants, being a new and brief method for computing their arithmetical values', *Proceedings of the Royal Society of London* 15 (1866), 150–5.

20. Letter from Spottiswoode to Dodgson, 8 February [1867], reproduced in *Mathematical Pamphlets*, pp. 160–1.

21. Spottiswoode (note 20).

22. Letter from Dodgson to Alexander Macmillan, 11 February 1867, quoted in *Lewis Carroll and the House of Macmillan* (ed. M. N. Cohen and A. Gandolfo), Cambridge University Press (1987), 49.

23. *Diaries*, 15 March 1867, Vol. 5, pp. 206–7.

24. *See Diaries*, Vol. 5, p. 370; note written on 23/24 November 1867.

25. Cohen and Gandolfo (note 22), p. 58, n.2.

26. *Diaries*, 12 December 1867, Vol. 5, p. 374, n.555.

27. C. G. J. Jacobi, 'De binis quibuslibet functionibus homogeneis secundi ordinis per substitutiones lineares in alias binas transformandis, quae solis quadratis variabilium constant; una cum variis theorematis de transformatione et determinatione integralium multiplicium', *Journal für die Reine und Angewandte Mathematik* 12 (1833), 1–69, on p. 9.

28. The general statement of Jacobi's theorem is: Let M be an $n \times n$ matrix, let $[M_{ij}]$ be an $m \times m$ minor of M, where $m < n$, let $[M'_{ij}]$ be the corresponding $m \times m$ minor of the adjugate M', and let $[M^{\star}_{ij}]$ be the complementary $(n - m) \times (n - m)$ minor of M. Then $\det[M'_{ij}] = (\det M)^{m-1} \cdot \det[M^{\star}_{ij}]$.

29. Fuller discussions of Dodgson's method of condensation may be found in Francine Abeles, 'Determinants and linear systems: Charles L. Dodgson's view', *British Journal for the History of Science* 19 (1986), 331–5; Adrian Rice and Eve Torrence, '"Shutting up like a telescope": Lewis Carroll's "curious" condensation method for evaluating determinants', *College Mathematics Journal* 38 (2007), 85–95; and Deanna Leggett, John Perry, and Eve Torrence, 'Computing determinants by double-crossing', *College Mathematics Journal* 42 (2011), 43–54.

30. *Determinants*, p. v.

31. In both his 1866 research paper and Appendix III of his 1867 book, Dodgson proved the validity of his condensation method. As we have seen, the method is grounded on an application of Jacobi's theorem, and in his proof Dodgson invoked this theorem to demonstrate the 3×3 case and the 4×4 case in suitable generality. But instead of providing a general demonstration of the $n \times n$ case, he disingenuously closed the proof with the abrupt words 'and similar proofs

might be given for larger Blocks' (p.124). Given that the rest of the book aimed to be as thoroughly rigorous as possible, this lapse of attention to detail is striking.

32. *Determinants*, p. 50.

33. *Determinants*, pp. iv–v.

34. *Determinants*, p. iv.

35. *Determinants*, p. iii.

36. Derek Hudson, *Lewis Carroll. An Illustrated Biography*, Constable (1976), 132.

37. Francine F. Abeles, 'Some Mathematical Letters of Charles L. Dodgson, 1866–1867', *Proc. Canadian Society for History and Philosophy of Mathematics* 10 (1997), 60–7, on p. 65.

38. Richard Baltzer, *Théorie et Applications des Déterminants* (transl. J. Hoüel), Mallet-Bachelier (1861), 47.

39. Charlie Lovett, *Lewis Carroll Among His Books: A Descriptive Catalogue of the Private Library of Charles L. Dodgson*, McFarland & Co. (2005), 49, 98–9, 235, 270, 316.

40. *Katz & Parshall*, pp. 414–15.

41. Isabella G. Bashmakova and Aleksei N. Rudakov, 'Algebra and Algebraic Number Theory', *Mathematics of the 19th Century: Mathematical Logic, Algebra, Number Theory, Probability Theory* (ed. A. N. Kolmogorov and A. P. Yushkevich), Birkhäuser (2001), 71. This is an English translation of the original Russian book, *Matematika XIX veka: Matematicheskaya logika, algebra, teoriya chisel, teoriya veroyatnostei*, published in 1978.

42. *Determinants*, p. 61.

43. *Mathematical Pamphlets*, p. 159.

44. *Muir*, Vol. 3, p. 24.

45. *Pall Mall Gazette*, 11 January 1868, p. 12.

46. *Pall Mall Gazette* (note 45), p. 12.

47. *The Educational Times*, 1 June 1868, p. 62.

48. *Educational Times* (note 47), p. 62.

49. Letter from Dodgson to Alexander Macmillan, 4 June 1868, quoted in Cohen and Gandolfo (note 22), p. 64.

50. *Educational Times* (note 47), p. 62.

51. *Pall Mall Gazette* (note 45), p. 12.

52. *Muir*, Vol. 3, p. 87.

53. P. S. Dwyer, *Linear Computations*, Wiley (1951); H. W. Turnbull, *The Theory of Determinants, Matrices, and Invariants*, 3rd edn, Dover, 1960.

54. David P. Robbins and Howard Rumsey Jr., 'Determinants and alternating sign matrices', *Advances in Mathematics* 62 (1986), 169–84.

55. Francine F. Abeles, 'Dodgson condensation: The historical and mathematical development of an experimental method', *Linear Algebra and its Applications* 429 (2008), 429–38, on p. 432.

56. See David M. Bressoud, *Proofs and Confirmations: The Story of the Alternating Sign Matrix Conjecture*, Cambridge University Press (1999).

57. See also Abeles (note 55).

CHAPTER 4: LOGIC

Dodgson's main logic works published in his lifetime were:

Lewis Carroll, *The Game of Logic*, Macmillan (1887). [*Game of Logic*]

Lewis Carroll, 'A logical paradox', *Mind* 3(11) (July 1894), 436–8.

Lewis Carroll, 'What the Tortoise said to Achilles', *Mind* 4(14) (April 1895), 278–80.

Lewis Carroll, *Symbolic Logic. Part 1: Elementary*, 4th edn, Macmillan (1897). [*Symbolic Logic*]

Symbolic Logic: Part I and *The Game of Logic* were published together as a Dover paperback in 1958.

There are two posthumous collections of Dodgson's rare or previously unpublished logic material:

William Warren Bartley III (ed.), *Lewis Carroll's Symbolic Logic*, C. N. Potter (1986).

Francine F. Abeles, *Logic Pamphlets*.

The main studies of Dodgson's logic include:

Francine F. Abeles, 'Lewis Carroll's formal logic' and 'Lewis Carroll's visual logic', *History and Philosophy of Logic* 26(1) (2005), 3–46, and 28(1) (2007), 1–17.

Amirouche Moktefi, 'Lewis Carroll's logic', in *British Logic in the Nineteenth Century* (ed. D. M. Gabbay and J. Woods), North-Holland (2008), 457–505.

Amirouche Moktefi and Francine F. Abeles (eds), '*What the Tortoise Said to Achilles*': Lewis Carroll's *Paradox of Inference*, special issue of: *The Carrollian: The Lewis Carroll Journal* 28 (2016).

For an overview of Victorian logic, see:

Dov M. Gabbay and John Woods (eds), *British Logic in the Nineteenth Century*, North-Holland (2008).

Ivor Grattan-Guinness, 'Victorian logic: From Whately to Russell', in *Victorian Britain*, pp. 359–74.

Notes

1. *Diaries*, 13 March and 6 September 1855, Vol. 1, pp. 74, 129.

2. For a logical reading of the *Alice* books, see the annotated editions by Peter Heath, *The Philosopher's Alice*, Academy editions (1974) and Martin Gardner, *The Annotated Alice: The Definitive Edition*, Penguin Books (2001), and also Bernard M. Patten, *The Logic of Alice*, Prometheus Books (2009).

3. *Determinants*, p. iii.

4. *Modern Rivals*, p. 7.

5. C. L. Dodgson, *Three Letters on Anti-Vaccination*, The Lewis Carroll Society (1976), 12–13.

6. Mark Richards, 'Charles Dodgson's work for God', in S. Lawrence and M. McCartney (eds), *Mathematicians and their Gods*, Oxford University Press (2015), 191–211.

7. *Collingwood*, p. 301.

8. Morton N. Cohen and Anita Gandolfo (eds), *Lewis Carroll and the House of Macmillan*, Cambridge University Press (1987), 319.

9. *Diaries*, 25 May 1876, Vol. 6, pp. 463–4.

10. *Diaries*, 29 March 1885, Vol. 8, p. 180.

11. *Diaries*, 25 July 1886, Vol. 8, p. 285.

12. See Clare Imholtz, 'The history of Lewis Carroll's *The Game of Logic*', *The Papers of the Bibliographical Society of America* 97(2) (2003), 183–213.

13. *The Literary World*, 16 April 1887, 121.

14. Cohen and Gandolfo (note 8), p. 290.

15. *Logic Pamphlets*, pp. 91–2.

16. *Bartley*, p. 47.

17. Hugh MacColl, 'Review of *Symbolic Logic*', *The Athenaeum*, 3599 (17 October 1896), 520–1. In a letter to Russell, on 17 May 1905, MacColl declared that his reading of Dodgson's *Symbolic Logic* 'rekindled the old fire which [he] thought extinct'. See Abeles and Moktefi, 'Hugh MacColl and Lewis Carroll: Crosscurrents in geometry and logic', *Philosophiae Scientiae* 15(1) (2011), 55–76.

18. *Symbolic Logic*, p. 185.

19. See Moktefi, 'Essay review: Lewis Carroll's Diaries: The Private Journals of Charles Lutwidge Dodgson (Lewis Carroll)/The Logic Pamphlets of Charles Lutwidge Dodgson and Related Pieces', *History and Philosophy of Logic* 39(2) (2018), 187–200.

20. *Symbolic Logic*, p. 184.

21. See Moktefi, 'Are other people's books difficult to read? The logic books in Lewis Carroll's private library', *Acta Baltica Historiae et Philosophiae Scientiarum* 5(1) (2017), 28–49.

22. George Boole, *An Investigation of the Laws of Thought*, Macmillan (1854). See Ivor Grattan-Guinness, 'George Boole, *An Investigation of the Laws of Thought on which are Founded the Mathematical Theory of Logic and Probabilities* (1854)', in *Landmark Writings in Western Mathematics 1640–1940*, Elsevier (2005), 470–9.

23. George Boole, *The Mathematical Analysis of Logic*, Macmillan (1847), 13.

24. John Venn, 'Symbolic Logic', *Princeton Review* 2 (1880), 248.

25. William Stanley Jevons, 'On the mechanical performance of logical inference', *Philosophical Transactions of the Royal Society of London* 160 (1870), 499.

26. Christine Ladd-Franklin, 'On the algebra of logic', *Studies in Logic*, Little, Brown, and Co. (1883), 17.

27. John Venn, 'On the diagrammatic and mechanical representation of propositions and reasonings', *London, Edinburgh, and Dublin Philosophical Magazine and Journal of Science* 10 (1880), 4–5.

28. The notion of universe of discourse seems to have been introduced by De Morgan, and then to have been used by Boole and his followers. See Ernest Coumet, 'The game of logic: A game of universes', *Lewis Carroll Observed* (ed. E. Guiliano), C. N. Potter (1976), 181–95.

29. See John N. Keynes, *Studies and Exercises in Formal Logic*, Macmillan (1906), 441–9.

30. *Symbolic Logic*, p. 172. See George Englebretsen and Nora Gilday, 'Lewis Carroll and the logic of negation', *Jabberwocky: The Journal of the Lewis Carroll Society* 5(2) (1976), 42–5.

31. *Symbolic Logic*, p. 172.

32. Traditional propositions of the form 'Some *x* are not *y*' were handled by Dodgson as 'Some *x* are not-*y*', so the negation of the copula is simply moved to the negation of the attribute.

33. *Symbolic Logic*, pp. 17–18.

34. *Symbolic Logic*, pp. 165–71.

35. *Game of Logic*, p. 19.

36. John Venn, *Symbolic Logic*, Macmillan (1894), 145.

37. *Symbolic Logic*, pp. 166–7.

38. See Moktefi, 'On the social utility of symbolic logic: Lewis Carroll against 'The Logicians', *Studia Metodologiczne*, 35 (2015), 133–50.

39. *Diaries*, 29 November 1884, Vol. 8, p. 155.

40. On the history of logic diagrams, see: Moktefi and Sun-Joo Shin, 'A history of logic diagrams', in Dov M. Gabbay, *Logic: A History of its Central Concepts* (ed. F. J. Pelletier and J. Woods), North-Holland (2012), 611–82.

41. See Anthony J. Macula, 'Lewis Carroll and the enumeration of minimal covers', *Mathematics Magazine* 68(4) (1995), 269–74. A rare instance of a problem privately worked out by Dodgson with a diagram for four terms is discussed in: Moktefi, 'Beyond syllogisms: Carroll's (marked) quadriliteral diagram', in A Moktefi and Sun-Joo Shin (eds), *Visual Reasoning with Diagrams*, Birkhäuser (2013), 55–71.

42. In *The Game of Logic* Dodgson used counters (red and grey) to indicate the state of the cells on a board.

43. *Symbolic Logic* (1897), p. 26.

44. *Symbolic Logic*, pp. 53–4.

45. *Symbolic Logic*, p. 72.

46. *Symbolic Logic*, p. 70.

47. *Symbolic Logic*, p. 183.

48. *Symbolic Logic*, p. 84. An early version of Dodgson's alogorithm appears in Moktefi, 'La théorie syllogistique de Lewis Carroll', *Cahiers Philosophiques de Strasbourg* 28(2) (2010), 207–24.

49. *Diaries*, 16 July 1894, Vol. 9, p. 155.

50. *Bartley*, p. 280.

51. C. Ladd-Franklin (1883). See Francine F. Abeles, 'Lewis Carroll's method of trees: its origin in *Studies in logic*', *Modern Logic* 1(1) (1990), 25–35.

52. See Mark R. Richards, 'Dodgson's methods of solving sorites problems: an investigation and a newly discovered process', *The Carrollian* 6 (2000), 35–40; Abeles (2005); Moktefi (2008).

53. See Moktefi and Abeles, 'The making of "What the Tortoise said to Achilles": Lewis Carroll's logical investigations toward a workable theory of hypotheticals', *The Carrollian* 28 (2016), 14–47.

54. Bertrand Russell, *A Fresh Look at Empiricism*, Routledge (1996), 525, 528.

55. *Diaries*, 5 February 1893, Vol. 9, pp. 52–3.

56. *Diaries*, 1 February 1894, Vol. 9, p. 124.

57. On the barbershop controversy see Amirouche Moktefi, 'Lewis Carroll and the British nineteenth-century logicians on the barber shop problem', *Proc. Canadian Society for the History and Philosophy of Mathematics* 20 (2007), 189–99.

58. 'A logical paradox', *Mind* 3(11) (1894), 438.

59. Bertrand Russell, *The Principles of Mathematics*, Cambridge University Press (1903), 18.

60. Russell (1903), p. 16. See also p. 35.

61. Gilbert Ryle, 'If, So, and Because', in Max Black (ed.), *Philosophical Analysis,* Prentice-Hall (1963), 306–7.

62. *Diaries*, 11–21 December 1894, Vol. 9, pp. 184–6.

63. *Bartley*, p. 448.

64. John Cook Wilson, *Statement and Inference*, Clarendon Press (1926), 637. On Dodgson's exchanges with Cook Wilson, see Mathieu Marion and Amirouche Moktefi, 'La logique symbolique en débat à Oxford à la fin du XIXe siècle: les disputes logiques de Lewis Carroll et John Cook Wilson', *Revue d'Histoire des Sciences* 67(2) (2014), 185–205.

65. Cohen and Gandolfo (1987), p. 323.

CHAPTER 5: VOTING

Dodgson's pamphlets on voting and related topics were:

C. L. Dodgson, *A Discussion of the Various Methods of Procedure in Conducting Elections*, E. B. Gardner, E. Pickard Hall, and J. H. Stacy, Printers to the University, Oxford (1873) [*Voting* 1];

C. L. Dodgson, *Suggestions As to the Best Method of Taking Votes, Where More Than Two Issues Are to Be Voted On*, E. Pickard Hall, and J. H. Stacy, Printers to the University, Oxford (1874) [*Voting* 2];

C. L. Dodgson, *A Method of Taking Votes on More Than Two Issues*, Clarendon Press (1876) [*Voting* 3];

C. L. Dodgson, *Lawn Tennis Tournaments: The True Method of Assigning Prizes with a Proof of the Fallacy of the Present Method*, Macmillan (1883) [*Lawn Tennis*];

C. L. Dodgson, *The Principles of Parliamentary Representation*, Harrison and Sons (1884) [*Voting* 4].

Dodgson's work on voting was first rediscovered by Duncan Black in his classic book,

Duncan Black, *The Theory of Committees and Elections,* Cambridge University Press (1958).

A revised and expanded edition (referred to below as *Black*) was edited by I. McLean, A. McMillan, and B. L. Monroe, Kluwer (1998), and contains a comprehensive bibliography of papers on Dodgson's voting theory, compiled by Monroe.

Dodgson's pamphlets have been reprinted with commentary in several places, most fully in

I. McLean and A. B. Urken, *Classics of Social Choice*, University of Michigan Press (1994) [*Social Choice*],

I. McLean, A. McMillan, and B. L. Monroe (eds), *A Mathematical Approach to Proportional Representation: Duncan Black on Lewis Carroll*, Kluwer (1996) [*PR*],

and in *Political Pamphlets*.

These works contain more contextual material than we could include here, as does

I. McLean, 'The strange history of social choice', *Handbook of Social Choice and Voting* (ed. J. C. Heckelman and N. R. Miller), Edward Elgar (2016), 15–34.

Notes

1. M. J. A. N. de Caritat, Marquis de Condorcet, *Essai sur l'Application de l'Analyse à la Probabilité des Décisions Rendues à la Pluralité des Voix*, Imprimerie Royale (1785).

2. In *Voting 3*, *Political Pamphlets*, p. 47, and *Social Choice*, p. 289.

3. The works of Dodgson's precursors are collected and translated with commentary in *Social Choice*, and in I. McLean and F. Hewitt, *Condorcet: Foundations of Social Choice and Political Theory*, Edward Elgar (1994). The two seminal works in the 20th century were *Black*, and K. J. Arrow's *Social Choice and Individual Values*, 2nd edn, Yale University Press (1963). For advanced students there are many guides to the modern literature of social choice; a widely used text is D. Mueller, *Public Choice III*, Cambridge University Press (2003).

4. In *Voting 1*, in *Political Pamphlets*, p. 35, and in *Social Choice*, pp. 279–96.

5. Ramon Llull's third text was discovered by Friedrich Pukelsheim and colleagues; their excellent site at www.math.uni-augsburg.de/emeriti/pukelsheim/llull/contains facsimiles of all three texts and translations into English, German, and modern Catalan.

6. Quoted in *Black*, p. 215.

7. Condorcet's tie-break procedure was first correctly described in H. P. Young, 'Condorcet's theory of voting', *American Political Science Review* 82(4) (1988), 1231–44.

8. In *Voting 1*, in *Political Pamphlets*, pp. 33–41, and in *Social Choice*, pp. 279–86.

9. The rediscoveries were by Iain McLean and Arnold B. Urken and their research teams, and appear in papers republished in *Social Choice*.

10. The search was verified (most recently in 2017) by kind courtesy of the Librarian, Christ Church.

11. *Black*, p. 228.

12. In *Voting 1*, in *Political Pamphlets*, p. 37, and in *Social Choice*, pp. 279–86.

13. *Black*, p. 236.

14. In *Voting 2*, in *Political Pamphlets*, pp. 43–5, and in *Social Choice*, pp. 287–8.

15. In *Voting* 3, in *Political Pamphlets*, p. 48, and in *Social Choice*, pp. 288–97.

16. In *Voting* 3, in *Political Pamphlets*, p. 50, and in *Social Choice*, pp. 288–97.

17. Kenneth O. May, 'A set of independent necessary and sufficient conditions for simple majority decision', *Econometrica* 20 (1952), 680–4.

18. For more context, including responses from tennis umpires to Dodgson's proposals, see *Political Pamphlets*, pp. 90–101.

19. See *Works*, p. 1082, and *Political Pamphlets*, p. 72.

20. In *Lawn Tennis* and in *Political Pamphlets*, p. 74.

21. R. D. McKelvey, 'Covering, dominance, and institution-free properties of social choice', *American Journal of Political Science* 30 (1986), 283–314.

22. I. McLean, 'Electoral Systems', in J. Fisher *et al.* (ed.), *The Routledge Handbook of Elections, Voting Behavior and Public Opinion*, Routledge (2018), 207–19.

23. J. S. Mill, *Considerations on Representative Government*, Parker, Son, and Bourn (1861), Chapter VII; Jenifer Hart, *Proportional Representation: Critics of the British Electoral System, 1820–1945*, Oxford University Press (1992).

24. The letters appear in *Political Pamphlets*, pp. 145–51.

25. Disraeli's surprise acceptances of hostile amendments shaped the reform more than once: see I. McLean, *Rational Choice and British Politics: An Analysis of Rhetoric and Manipulation from Peel to Blair*, Oxford University Press (2001), 61–85.

26. 3rd Marquess of Salisbury, 'The value of redistribution: a note on electoral statistics', *National Review* 4 (1882), 145–62, reprinted in *Political Pamphlets*, pp. 227–47.

27. This is in the estimation of most qualified observers. The unique 1851 census of religious attendance was not extended to Ireland, so no authoritative figures are available.

28. Sir John Lubbock. *Representation*, Swan Sonnenschein (1885), 20–1.

29. For evidence for the next three quotations see *Diaries*, 17 May 1882, Vol. 7, pp. 427–8, and 3–4 June 1884, Vol. 8, p. 119, and *Letters*: cited in *PR*, p. xxxii (emphases in originals).

30. J. Cornford, 'The transformation of Conservatism in the late nineteenth century', *Victorian Studies* 7 (1963), 35–66, and McLean (note 25), pp. 79–85.

31. In *Voting* 4, in *Political Pamphlets*, p. 194, and in *Social Choice*, pp. 299–320.

32. G. Cox and E. Niou, 'Seat bonuses under the Single Nontransferable Vote system: evidence from Japan and Taiwan', *Comparative Politics* 26 (1994), 221–36.

33. See, for example, C. L. Dodgson, *The Vision of the Three T's*, 2nd edn, James Parker (1873), part quoted and discussed in *Black*, pp. 232–3. It may be read in *Works*, pp. 1036–53, and in facsimile at en.wikisource.org/wiki/The_Vision_of_the_Three_T%27s.

CHAPTER 6: RECREATIONAL MATHEMATICS

Further information on Dodgson's recreational mathematics can be found in
Christopher Morgan (ed.), *The Pamphlets of Lewis Carroll, 5: Games, Puzzles, & Related Pieces*, Lewis Carroll Society of North America (2015).

Dodgson's puzzles and paradoxes are discussed at length in

Edward Wakeling, *Lewis Carroll's Games and Puzzles* and *Rediscovered Lewis Carroll Puzzles*, Dover Publications (1992, 1995)

and also in

John Fisher, *The Magic of Lewis Carroll*, Simon and Schuster (1973),

Martin Gardner, *The Universe in a Handkerchief*, Springer-Verlag (1996),

and in Chapter 7 of *Numberland*.

Notes

1. *Diaries*, 30 August 1855, Vol. 1, p. 126.

2. *Diaries*, 5 and 8 February 1856, Vol. 2, pp. 34, 36.

3. Eight Dodgson family magazines for internal consumption were compiled between 1845 and 1862. Of these, only four survive. The original *The Rectory Umbrella* is at Harvard University. The magazine has been edited with the last magazine *Mischmasch* by Florence Miller and was published by Cassell in 1932; for 'Difficulties, No. 1', see p. 31.

4. *Diaries*, 23 February 1857, Vol. 3, p. 29.

5. The Eastern Telegraph Company was formed in 1857 to link Malta into the French and Italian land system of cables; this was achieved via a cable from Sardinia. It continued to lay cables around the globe. Dodgson's reply came from the Traffic General Controller, W. T. Ansell, and is dated 25 April 1885. The manuscript is in the Warren Weaver Collection, University of Texas at Austin, *Lewis Carroll at Texas*, item 694, on p. 182.

6. For more on Pillow-Problems, see Robin Wilson, 'The Pillow-Problems of Charles L. Dodgson', *Proc. Recreational Mathematics Colloquium V (Gathering for Gardner Europe)* (ed. J. N. Silva), Associação Ludus, Lisbon (2017), 147–56.

7. *Diaries*, 12 March 1856, Vol. 2, p. 50.

8. Dodgson's letter is quoted in full in *Diaries*, 15 November 1866, Vol. 5, pp. 183–4.

9. *Diaries*, 16 January 1858, Vol. 3, pp. 151–2.

10. *Diaries*, 29 February 1856, Vol. 2, pp. 44–5; Edward Pember was a Student at Christ Church from 1854 to 1861.

11. The ten 'counters-in-bags' problems are Questions 5, 16, 19, 23, 27, 38, 41, 50, 66, and 72.

12. The puzzles were in the possession of Dr Francis Vernon Price (1913–2012), Fellow and Tutor in Physics, Worcester College, Oxford, and grandson of Prof. Bartholomew Price. Some years later this collection of manuscripts was acquired by the author.

13. This letter is in the author's collection. Edward Sampson took over the mathematical lectureship at Christ Church after Dodgson retired at the end of 1881.

14. David Singmaster of South Bank University, London, gave some background to this problem, in 'Coconuts: the history and solutions of a classic Diophantine problem', *Ganita Bhāratī* (Bulletin of the Indian Society for the History of Mathematics), 19 (1997), 35–51.

15. Henry Dudeney's *The Canterbury Puzzles and Other Curious Problems* was first published in 1907, and a revised edition appeared in 1919. It is so named because it opens with puzzles that are based on characters from Geoffrey Chaucer's *The Canterbury Tales*.

16. These diary entries are from *Diaries*, 17, 22, and 23 August 1879, Vol. 7, pp. 137, 202.

17. Dodgson's remark was quoted by Thomas Strong, Bishop of Oxford and a former colleague of Dodgson, in 'Mr. Dodgson: Lewis Carroll at Oxford' and was published in *The Times* on 27 January 1932 on pp. 11–12.

18. Dodgson named each part of *A Tangled Tale* as a 'knot' to signify the difficulty of the problems that it featured. Knots also feature n Chapter 3 of *Alice's Adventures in Wonderland*; when the Mouse responds to Alice's comment on his story by crying, 'I had not!', Alice replies, 'A knot! Oh, do let me help to undo it!'.

19. *Diaries*, 3 July 1885, Vol. 8, p. 218.

20. This is among Dodgson's mathematical papers in the Parrish Collection at Princeton University.

21. *Diaries*, 16 February 1890, Vol. 8, pp. 504–5.

22. The four-colour problem was due to Francis Guthrie, a former student of Augustus De Morgan. De Morgan was intrigued by the problem and posed it in a letter to Sir William Rowan Hamilton on 23 October 1852, but it remained unanswered for many years. It was finally answered in the affirmative by Kenneth Appel and Wolfgang Haken of the University of Illinois in 1976; see Robin Wilson, *Four Colors Suffice* (revised edn), Princeton University Press (2013).

23. *Collingwood*, pp. 370–1.

24. For the diary entries on *Alice's Puzzle-Book*, see *Diaries*, 1 March 1875, Vol. 6, pp. 381–2, 23 November 1881, Vol. 7, p. 380, 29 March 1885, Vol. 8, p. 181, and 12 November 1897, Vol. 9, p. 351.

25. Dr Richard Grey (1694–1771) published *Memoria Technica, or Method of Artificial Memory* in 1730. Dodgson probably consulted a revised edition, published in 1851 by J. Vincent of Oxford.

26. *Diaries*, 27/28 October 1875, Vol. 6, p. 428.

27. *Diaries*, 21 March 1878, Vol. 7, p. 102.

28. *Diaries*, 27/28 September 1897, Vol. 9, pp. 341–2.

29. A copy of this galley proof is in the Parrish Collection at Princeton University.

30. By his own admission, 'Abridged long division' used ideas put forth by others that he improved on, particularly an accuracy test. In its emphasis on minimizing the number of steps in an algorithm, this paper has implications for modern computing.

31. *Diaries*, 8 March 1887, Vol. 8, p. 323; in 1973 the mathematician John Conway presented a different method for finding the day of the week for any given date.

32. This manuscript is in the Parrish Collection at Princeton University.

33. For the diary entries on number-guessing puzzles, see *Diaries*, 6 January 1895, 3 February 1896, and 15 February and 15 March 1897, Vol. 9, pp. 187, 237, 239, and 299.

34. This manuscript is in the Warren Weaver Collection, Harry Ransom Center, University of Texas at Austin.

35. The manuscript is published in *Mathematical Pamphlets*, p. 289.

36. *Diaries*, 6 October 1897, Vol. 9, p. 343. Mr C. and Mr T. and MOWS were puzzle stories. In the latter, the letters gradually transformed into a pig; the meaning of the letters is unknown.

37. Such cyclical numbers were known to Dodgson as *circulating decimals*. In 'The Offer of the Clarendon Trustees' (1868), Dodgson proposed that, since laboratories were being built for the other sciences, mathematics ought to have its own labs. Among these, he suggested 'A large room, which might be darkened, and fitted with a magic lantern, for the purpose of exhibiting Circulating Decimals in the act of circulation'.

38. In a similar way, if a three-digit number using different digits is chosen and the same procedure is carried out, then the result is always 1089. It is not known whether these puzzles originated with Dodgson.

CHAPTER 7: MATHEMATICAL LEGACY

Further reading for this chapter can be found in the Further reading sections of earlier chapters

1. See also *Mathematical Pamphlets*, pp. 27–34.

2. *Modern Rivals*, p. vii.

3. Richard A. Proctor, 'Chats about geometrical measurement', *Knowledge* 6 (24 October 1884), 337–9. See also Francine F. Abeles, 'How did Charles L. Dodgson view the non-Euclidean geometries?', *The Carrollian: The Lewis Carroll Journal* 23 (2012), 40–3.

4. Charles L. Dodgson, *Supplement to Euclid and his Modern Rivals*, Macmillan (1885), 351.

5. *Diaries*, 24–26 March 1888, Vol. 8, pp. 387–9.

6. See *Diaries*, 8 April 1888, Vol. 8, p. 390, and *Parallels*, first edn., Appendix IV.

7. *Parallels* (first edn.), p. xiv.

8. See Francine F. Abeles, 'Dodgson's engagement with non-finite processes, 1885–1895', *The Carrollian: The Lewis Carroll Journal* No. 30 (October 2017), 18–32.

9. See Francine F. Abeles and Amirouche Moktefi, 'Hugh MacColl and Lewis Carroll: Crosscurrents in geometry and logic', *Philosophiae Scientiae* 15(1) (2011), 55–76.

10. Hugh MacColl, 'Review of C. L. Dodgson's *Curiosa Mathematica*. Part I: *A New Theory of Parallels*' (first edn), *The Athenaeum* 3183 (27 October 1888), 557.

11. *Parallels* (second edn), p. xi.

12. Hugh MacColl, 'Review of C. L. Dodgson's *Curiosa Mathematica*. Part I: *A New Theory of Parallels*' (third edn), *The Athenaeum* 3328 (8 August 1891), 196–7.

13. *Parallels* (note 7), p. xv.

14. Hugh MacColl, 'Linguistic misunderstandings. Part I', *Mind* 19 (1910), 186–99, on p. 193.

15. MacColl (note 14), p. 193.

16. MacColl (note 14), p. 196.

17. *Parallels* (third edn), p. 42.

18. Abeles and Moktefi (note 9).

19. *Parallels* (note 7), p. 51.

20. *Determinants*, pp. i, iv, v.

21. For more details, see Francine F. Abeles, 'Dodgson condensation, the historical and mathematical development of an experimental method', *Linear Algebra and its Applications* 429 (2008), 429–38, and Eugene Seneta, 'Lewis Carroll as a probabilist and mathematician', *Math. Scientist* 9 (1984), 79–84.

22. For more details, see David P. Robbins and Howard Rumsey, Jr., 'Determinants and alternating sign matrices', *Advances in Math.* 62 (1986), 169–84, and David M. Bressoud, *Proofs and Confirmations. The Story of the Alternating Sign Matrix Conjecture*, Math. Assoc. of America (1999).

23. *Determinants* (note 20), pp. 122–4. For a modern approach, see Adrian Rice and Eve Torrance, ' "Shutting up like a telescope": Lewis Carroll's "curious" condensation method for evaluating determinants', *College Mathematical Journal* 38 (2007), 85–95.

24. David M. Bressoud and James Propp, 'How the alternating sign matrix conjecture was solved', *Notices of the American Mathematical Society* 46 (1999), 637–46, on p. 638.

25. See Francine F. Abeles, 'Determinants and linear systems: Charles L. Dodgson's view', *British J. History of Science* 19 (1986), 331–5, and Tewodros Amdeberhan and Shalosh B. Ekhad, 'A condensed condensation proof of a determinant evaluation conjectured by Greg Kuperberg and Jim Propp', *J. Combin. Theory (A)* 78 (1997), 169–70.

26. Doron Zeilberger, 'Dodgson's determinant evaluation rule proved by two-timing men and women', *Electron. J. Combin.* 4 (1997), R22.

27. See Thomas Muir, 'The law of extensible minors in determinants', *Trans. Royal Society of Edinburgh* (1883), 1–4, and Adam Berliner and Richard A. Brualdi, 'A combinatorial proof of the Dodgson/Muir determinantal identity', *Internat. J. Information and Systems Sci.* 4(1) (2008), 1–7.

28. See Percy A. MacMahon, *Combinatory Analysis*, Cambridge University Press (1915–16), Section X, Chap. 1, reprinted by Chelsea (1960), and Doron Zeilberger, 'Reverend Charles to the aid of Major Percy and Fields–Medalist Enrico', *American Math. Monthly* 103 (1996), 501–2.

29. Tewodros Amdeberhan and Doron Zeilberger, 'Determinants through the looking glass', *Advances in Applied Math.* 27 (2001), 225–30.

30. Doron Zeilberger, 'Liebe Opa Paul, ich bin ein experimental Scientist!' *Advances in Applied Math.* 31 (2003), 532–43.

31. David P. Robbins, 'A conjecture about Dodgson condensation', *Advances in Applied Math.* 34 (2005), 654–8, on p. 657.

32. I. M. Gelfand and V. S. Retakh, 'Determinants of matrices over noncommutative rings', *Funct. Anal. Appl.* 25 (1991), 91–102.

33. See Israel Gelfand, Sergei Gelfand, Vladimir Retakh, and Robert Lee Wilson, 'Quasideterminants', *Advances in Math.* 193 (2005), 56–141, and I. M. Gelfand and V. S. Retakh, 'Quasideterminants, *Selecta Math.* 3 (1997), 517–46.

34. *Logic Pamphlets*, p. 3.

35. *Logic Pamphlets*, p. 6.

36. William Warren Bartley, III (ed.), *Lewis Carroll's Symbolic Logic*, Clarkson N. Potter (1977), 477ff.

37. Bartley (note 36), p. 46.

38. [Review of *Symbolic Logic, Part I*], *The Educational Times*, 1 July 1896, 316.

39. Bartley (note 36), pp. 499–500.

40. Russell (1903), p. 18.

41. See Bartley (note 40), p. 32.

42. Irving H. Anellis, 'From semantic tableaux to Smullyan trees: a history of the development of the falsifiability tree method', *Modern Logic* 1 (1990), 36–69. See also Francine F. Abeles's articles, 'Lewis Carroll's formal logic', *History and Philosophy of Logic* 26 (2005), 33–46, and 'Toward a visual proof system: Lewis Carroll's method of trees', *Logica Universalis* 6 (2012), 521–34. Jaakko Hintilla was a Finnish logician and Raymond M. Smullyan was an American mathematician, philosopher, magician, and concert pianist who created many logic puzzles.

43. *The Athenaeum* (1896), 520–1, and *The Educational Times*, 1 July 1896, 316.

44. Lewis Carroll, *The Mathematical Recreations of Lewis Carroll*; reprint of *The Game of Logic* and *Symbolic Logic, Part I*, Dover (1958), 81.

45. A. W. F. Edwards, *Cogwheels of the Mind: The Story of Venn Diagrams*, Johns Hopkins University Press (2004), 27, and Anthony J. Macula, 'Lewis Carroll and the enumeration of minimal covers', *Math. Magazine* 69 (1995), 269–74.

46. Carroll (note 44), pp. 182–3.

47. The manuscript is in the Lindseth Collection, L188. For a comprehensive view of Dodgson's work in logic, see Amirouche Moktefi, 'Lewis Carroll's logic', *Handbook of the History of Logic 4: British Logic in the Nineteenth Century* (ed. D. M. Gabbay and J. Woods), Elsevier North-Holland (2008), 457–505.

48. See Michael Astroh, Ivor Grattan-Guinness, and Stephen Read, 'A survey of the life of Hugh McColl (1837–1909)', *History and Philosophy of Logic* 22 (2) (2001), 81–98.

49. Charles L. Dodgson, 'A Logical Paradox', *Mind* 3 (new series) (1894), 436–8; 'What the Tortoise Said to Achilles', *Mind* 4 (new series) (1895), 278–80. See also the section on hypotheticals in Chapter 4, and Amirouche Moktefi and Francine F. Abeles (eds.), 'What the Tortoise Said to Achilles', Lewis Carroll's Paradox of Inference', *The Carrollian: The Lewis Carroll Journal* 28 (November 2016).

50. E. Lusk and R. Overbeek, 'Non-Horn problems', *J. Automated Reasoning* 1 (1985), 103–14; see also A. G. Cohn, 'On the appearance of sorted literals: a non-substitutional framework for hybrid reasoning', *Proc. First Internat. Conf. on Principles of Representation and Reasoning,*

M. Kaufman (1989), 55–66. For a wider perspective, see Francine F. Abeles, 'Logic and Lewis Carroll', *Nature* 527 (19 November 2015), 302–4.

51. Duncan Black, *The Theory of Committees and Elections*, Cambridge University Press, 1958. See also I. McLean, A. McMillan, and B. L. Monroe (eds), *A Mathematical Approach to Proportional Representation: Duncan Black on Lewis Carroll*, Kluwer (1996).

52. John von Neumann, 'Zur Theorie der Gesellschaftsspiele', *Math. Annalen* 100 (1928), 295–300.

53. C. L. Dodgson, *A Method of Taking Votes on More Than Two Issues*, Clarendon Press (1876), 18–19. For complete details see Francine Abeles, 'The mathematical–political papers of C. L. Dodgson', *Lewis Carroll: A Celebration* (ed. E. Guiliano), Clarkson N. Potter (1982), 195–210.

54. *Diaries*, 22 September 1857, Vol. 3, pp. 111–13.

55. C. L. Dodgson, *The Principles of Parliamentary Representation*, Harrison and Sons (1884).

56. E. V. Huntington, 'The mathematical theory of apportionment of representatives', *Proc. National Academy of Sciences* 4 (1921), 123–7.

57. Donald Knuth, *The Art of Computer Programming*, Vol. II, Prentice Hall (1973), 211.

58. See, for example, E. Hemaspaandra and L. A. Hemaspaandra, 'Computational politics: Electoral systems', *Mathematical Foundations of Computer Science* (ed. M. Nielsen and B. Rovan), Springer (2000), 64–83, T. C. Ratliff, 'Some startling inconsistencies when electing committees', *Social Choice and Welfare* 21 (2003), 433–54, and J. Rothe *et al.*, 'Exact complexity of the winner problem for young elections', *Theory of Computing Systems* 36 (2003), 375–86.

59. In this section we rely extensively on the writings of Eugene Seneta, who has explored Dodgson's work on probability in great detail; see also *Mathematical Pamphlets*, pp. 201–8.

60. Eugene Seneta, 'Lewis Carroll, Boole's Inequality and statistical inference', Lewis Carroll: Man of Science, *The Carrollian: The Lewis Carroll Journal* 30 (October 2017), 33–44, on pp. 36–8.

61. *Diaries*, 12 March 1856, Vol. 2, p. 50.

62. Lewis Carroll, 'The science of betting', *The Pall Mall Gazette* (19 November 1866), 3.

63. See *Diaries*, Vol. 6, p. 453.

64. See Eugene Seneta's articles 'Lewis Carroll as a probabilist and mathematician', *Math. Scientist* 9 (1984), 79–94, and 'Lewis Carroll's "Pillow Problems" on the 1993 centenary', *Statistical Science* 8 (1993), 180–6.

65. Charles L. Dodgson, 'Note on Question 7695', 'Something or Nothing', and 'Question 9588', *The Educational Times* XXXVIII (1 May 1885), 183, and XLI (1 June 1888), 245, 247.

66. There are several references to Cook Wilson in Dodgson's diaries: 27 February and 25 April 1885, 23 November 1886, 30 May 1891, 21 January and 5 February 1893; 1 February and 31 March 1894, and 15 February 1897, Vol. 8, pp 169, 192, 306, and 559, and Vol 9, pp. 51, 52, 124, 137, and 295.

67. John Cook Wilson, 'Inverse or "a posteriori" Probability', *Nature* 63 (1900), 154–6.

68. For an introduction to Dodgson's work on ciphers, see *Mathematical Pamphlets*, pp. 325–35, and for more information on cipher systems in general, see David Kahn, *The Codebreakers*, Macmillan (1967). The analysis of the alphabet, telegraph, key-vowel, and matrix ciphers is in

this section see *Diaries*, Vol. 3, pp. 159–64, and *Mathematical Pamphlets*, pp. 336–45. The analysis of the alphabet, telegraph key-vowel and matrix ciphers is joint work with Stanley H. Lipson.

69. The telegraph-cipher card and description can be found in *Mathematical Pamphlets*, pp. 324 and 344–5.

70. See F. Beaufort, *Cryptography. A System of Secret Writing by the late Admiral Sir Francis Beaufort, K.C.B., adapted for telegrams and postcards* (Card), Edward Sanford.

71. For a more complete description, see A. Kerckhoffs, *La Cryptographie Militaire*, L. Baudoin et Cie (1883).

72. For more details, see Francine F. Abeles, 'A triad of mathematics popularizers: Martin Gardner, Richard Proctor, and Charles Dodgson', in Mark Burstein (ed.), *A Bouquet for the Gardener: Martin Gardner Remembered*, Lewis Carroll Society of North America (2011), 150–9.

73. In a recent paper, 'Dodgson's polynomial identities', arXiv:1810.06220, Marcel Golz provides a new combinatorial interpretation and generalization of Dodgson polynomials, leading to two new identities that can simplify the parametric integrand for quantum electrodynamics.

NOTES ON CONTRIBUTORS

Francine F. Abeles is a Professor Emerita (formerly a Distinguished Professor of Mathematics and Computer Science) at Kean University in Union, New Jersey, USA. She has edited three volumes in the Pamphlets of Lewis Carroll series (Mathematical, Political, and Logic Pamphlets) for the Lewis Carroll Society of North America (LCSNA), and with Amirouche Moktefi she has recently co-edited 'What the Tortoise Said to Achilles'. Lewis Carroll's Paradox of Inference. For the LCSNA, she is a member of the Publications Committee, and managing trustee of the Morton N. Cohen Publications Trust. She is the author of nearly one hundred papers, many related to Charles Dodgson, on topics in geometry, number theory, voting theory, linear algebra, logic, and their history, and regularly reviews articles for *MathSciNet*,

Iain McLean is a Senior Research Fellow at Nuffield College, Oxford University, and a Fellow of the British Academy and the Royal Society of Edinburgh. He has been working on the lost history of the mathematics of voting for many years, being responsible (with co-authors) for the rediscovery of Ramon Llull and Nicolaus Cusanus in this context. He has also conserved the papers of Duncan Black, the first modern scholar to understand Dodgson's contributions to voting theory. Relevant publications include *Classics of Social Choice* (edited and translated with Arnold B. Urken) and *Duncan Black on Lewis Carroll* (edited with Alistair McMillan and Burt L. Monroe).

Amirouche Moktefi is a Lecturer in Philosophy at Tallinn University of Technology, Estonia. He is a member of the *Ragnar Nurkse Department of Innovation and Governance*, Estonia, and is associated with the *Archives Henri Poincaré – Philosophy and Researches on Sciences and Technologies* in France. His areas of interest include the history of mathematics and logic and the philosophy of visual reasoning. He has published extensively on Dodgson's mathematics and logic, and has recently co-edited two special issues devoted to Dodgson's mathematics of *The Carrollian: The Lewis Carroll Journal*: 'What the Tortoise Said to Achilles': Lewis Carroll's Paradox of Inference (with Francine F. Abeles) and *Lewis Carroll: Man of Science* (with Mark Richards and Robin Wilson).

Adrian Rice is a Professor of Mathematics at Randolph-Macon College in Virginia, USA, where his research focuses on the history of mathematics – particularly mathematical developments in 19th- and early 20th-century Britain. His publications include *Mathematics Unbound: The Evolution of an International Mathematical Research Community, 1800–1945* (edited with Karen

Hunger Parshall), the London Mathematical Society's *Book of Presidents, 1865–1965* (with Susan Oakes and Alan Pears), *Mathematics in Victorian Britain* (edited with Raymond Flood and Robin Wilson), and *Ada Lovelace: The Making of a Computer Scientist* (with Christopher Hollings and Ursula Martin). He is a three-time recipient of awards for outstanding expository writing from the Mathematical Association of America.

Mark R. Richards is a retired IT consultant, independent scholar and Carroll collector. He has studied Charles Dodgson's mathematics and logic for over forty years, and has served as Chair of the Lewis Carroll Society and is a former Editor of the scholarly journal, *The Carrollian*. His chapter on Dodgson in Oxford University Press's *Mathematicians and their Gods* makes the case that it is in the interrelationship between Dodgson's mathematics and his Christian faith that we can find a true understanding of the man. He is currently developing *lewiscarrollresources.net*, a wide-ranging collection of online facilities and databases to promote Carroll scholarship and to assist future researchers.

Edward Wakeling is a mathematician by profession and a long-standing member of the Lewis Carroll Society in which he has held the posts of Secretary, Treasurer, and Chairman. For a time he was an educational adviser becoming Bedfordshire's Inspector for Mathematics, now retired. He has written widely on Dodgson over the last four decades, and among his publications is the first unabridged edition of *Lewis Carroll's Diaries* in ten volumes (Lewis Carroll Society, 1993–2007). More recently he has produced *The Photographs of Lewis Carroll, A Catalogue Raisonné* and *Lewis Carroll, The Man and his Circle*. As a recognised Carrollian scholar and collector, he is frequently called upon to contribute to conferences, exhibitions, and television programmes around the world.

Robin Wilson is an Emeritus Professor of Pure Mathematics at the Open University, Emeritus Professor of Geometry at Gresham College, London, and a former Fellow of Keble College, Oxford University. He is currently a Visiting Professor at the London School of Economics. A former President of the British Society for the History of Mathematics, he has written and edited over forty books on the history of mathematics, including *Lewis Carroll in Numberland*, and also on graph theory and combinatorics. Involved with the popularization and communication of mathematics and its history, he has been awarded prizes by the Mathematical Association of America for his 'outstanding expository writing', and the Stanton Medal by the Institute of Combinatorics and its Applications for outreach activities in combinatorial mathematics.

INDEX